To Roberta and Jeffrey, Debbie and Mark, David and Russell.

THE BUILD-IT BOOK OF
ELECTRONIC PROJECTS

BY RUDOLF F. GRAF & GEORGE J. WHALEN

TAB BOOKS Inc.

BLUE RIDGE SUMMIT, PA 17214

FIRST EDITION

SECOND PRINTING

Printed in the United States of America

Reproduction or publication of the content in any manner, without express permission of the publisher, is prohibited. No liability is assumed with respect to the use of the information herein.

Library of Congress Cataloging in Publication Data

Graf, Rudolf F.
    The build-it book of electronic projects.

    Includes index.
    1. Electronics—Amateurs' manuals. I. Whalen,
George J. II. Title.
TK9965.G67   1983   621.381   83-4896
ISBN 0-8306-0498-7
ISBN 0-8306-1498-2 (pbk.)

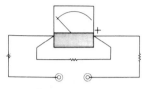

Contents

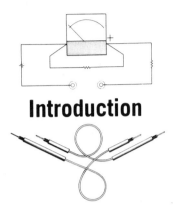

Introduction

If you enjoy the challenge and creativity of building, you'll want to construct the electronic projects described in this book. Electronic project building is not only fun, it's also one of the best ways to learn about electronics. In addition, when you're done, you'll experience the self-satisfaction of using your completed project.

This collection of electronic projects includes practical devices for use in your home and car; devices designed specifically to protect your family, home, automobile, and other possessions; and devices designed just for fun and game applications. A telephone-conversation amplifier, automatic burglar alarm, audible tachometer, electronic football game, visible and audible metronome, metal detector, medicine cabinet alarm, power-line monitor, electronic temperature controller . . . these are just a few of the 50 projects presented.

All of the projects involve relatively simple circuits and use the latest solid-state devices and integrated circuits. All have been built and tested to ensure optimum performance. Regardless of how you use these projects, you'll find them interesting and worthwhile. They'll provide you with hours of building enjoyment and help you appreciate the many ways electronics can be put to practical use.

Special thanks go to the following people whose generous assistance in the preparation of this book we gratefully acknowledge:

Barbara Bruno and Mrs. John J. Dillon who helped in the typing of the manuscript.

Michael Zarembski and Leonard Haber for meritorious service in the photo lab.

The personnel of Howard W. Sams & Co., Inc. and *Radio-Electronics* and *Mechanix Illustrated* magazines and especially thanks to *Popular Mechanics* and *Popular Science* in whose fine magazines some of these projects originally appeared.

Rachmiel Block and Charles Cimilluca for their thoughtful and worthwhile comments on this material.

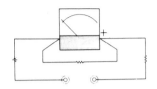

Home Electronics

As electronics grows more important in our jobs, our cars, our entertainment, and our future, it's not surprising that it is finding a growing role in our lives at home, too. Television, stereo, lamp dimmers, microwave ovens, and calculators are just a few of the home electronic devices that are helping make our lives easier around the home. But, there are countless other jobs that electronics can do around the household. This chapter presents a sampling of unusual projects that can put electronics to work for you in your home.

Want an automatic telephone secretary that transcribes both sides of a telephone conversation? Want a device to tell you when to water your lawn or garden? Could you use a remote phone bell in the backyard that you plug in when you want to hear a call? These are just some of the projects that await you.

Spend about one dollar for parts and in a half hour you can whip up a useful little gadget that converts your tape recorder into a *completely automatic* telephone conversation recording instrument that *needs no external power source* (Fig. 1-1).

As soon as anyone picks up the phone (in response to an incoming call *or* to dial out), your tape recorder automatically turns on to record every word of *both* sides of the conversation. Hang up the phone and the tape recorder shuts off automatically. The gadget itself uses no batteries and has no noticeable effect on the normal operation of your telephone. Simply connect two wires to a telephone jack—or anywhere else across your two telephone wires— and you can make your tape recorder serve as an unfailing memory for *all* telephone conversations. The TCTS is external to the tape recorder and requires absolutely no modification of the machine.

The unit plugs into the tape recorder jack where the microphone is normally plugged in. Look closely and you will see that this plug is in effect two plugs with four wires going into the mike's housing. One pair of wires goes to the microphone itself, and the other two go to the on-off switch. You can use such a plug and the wire from a microphone that fits your machine, or get two suitable separate plugs. We used the wire and plug from a mike. It's neater and easier.

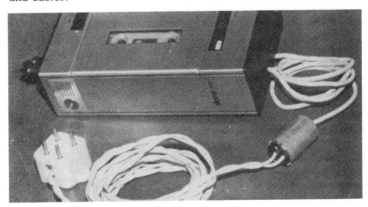

Fig. 1-1. Telephone-Controlled Tape Starter

How It Works

Refer to Fig. 1-2. With the switch terminals of your tape recorder "open," you can measure a voltage across them that is

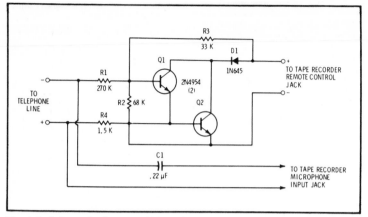

Fig. 1-2. Schematic diagram of Tape Starter.

equal to the dc voltage used to operate the machine. This is usually around 6 volts. If we apply this readily available voltage to a pair of Darlington-connected transistors, Q1 and Q2, they will turn on and start the tape recorder. To turn the transistors off, and thereby stop the machine, we have to apply a negative voltage to the base of Q1. We get that voltage from the phone line.

When the telephone receiver is on the hook, there is typically about 50 volts dc across the phone line. This voltage is divided across R1, R2, and R4 in such a way that the base of Q1 is sufficiently negative to keep the tape recorder off. When the phone's receiver is picked up, the voltage on the telephone line drops to about 5 volts, which leaves insufficient negative voltage on the base of Q1 to keep it cut off, so the tape recorder starts and begins to record.

When the phone rings and the TCTS is connected, the recorder turns on with every ring. So you can even tell how long it took to get to the phone, or if no one picks it up, you know that a call came in, the number of times it rang, and that no one answered. Should you have an answering service and *they* pick up the phone, you will also record that conversation.

Construction

We found that a fluorescent starter housing (Fig. 1-3) was convenient and of sufficient size to house the two transistors, the diode, the four resistors, and the coupling capacitor. (You can duplicate our tight assembly, or use any other suitable container.)

Remove and discard the starter element, but save the Bakelite base for use as a convenient terminal board for all the components.

4

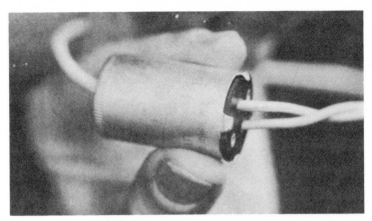

Fig. 1-3. Fluorescent starter housing.

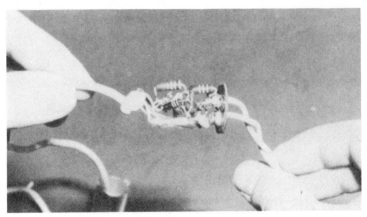

Fig. 1-4. Point-to-point wiring is used.

We used point-to-point wiring for minimum size, as shown in the photo (Fig. 1-4). Invert the base so that the brass terminals are on the inside. That makes wiring easier. A rubber grommet in the bottom of the aluminum cover protects the wires to the recorder. The wires to the phone jack can be directly soldered to the terminals, or they can be inserted through the two extra holes drilled in the cover.

Installation and Use

Plug the TCTS into the proper tape recorder jacks and set the machine to playback. With the TCTS not connected to the phone line, the tape recorder should start. If it doesn't, then the wires that go to the tape recorder remote control jack are reversed. Switch them.

Now that the machine is playing with the TCTS plugged in, connect the two wires to the phone line. With the phone on the hook, the tape recorder should stop. If it doesn't, reverse the two wires leading to the phone line. Now that the tape recorder has stopped, lift the phone to check that the recorder will start.

To set up for automatic recording, press the forward and record buttons on the tape recorder. Lift the receiver and adjust the tape recorder for proper recording level of both sides of the conversation. We found that for optimum recording quality, it is best to speak *over* the phone's mike rather than directly into it, or else you will over-record, and the playback will be distorted and difficult to understand.

The TCTS works equally well with a recorder that is operated from internal batteries or one that is operated from an ac adapter. Since the TCTS places a light load (about 60 microamperes) on the recorder's power supply, it is best to operate from the ac line, unless the unit will be "on call" for only a few days, or an ac source is not conveniently available. A complete parts list for the TCTS is given in Table 1-1.

Table 1-1. Telephone-Controlled Tape Starter Parts List

R1—270k, ¼-W, ±10% resistor
R2—68k, ¼-W, ±10% resistor
R3—33k, ¼-W, ±10% resistor
R4—1.5k, ¼-W, ±10% resistor
Q1, Q2—2N4954 transistor (Radio Shack 276-2009, or equiv.)
D1—1N645 diode (Radio Shack 276-1104, or equiv.)
C1—0.22-μF, 50-V dipped solid tantalum capacitor
Misc—telephone plug, fluorescent starter housing, wire, solder

2. SOIL MOISTURE CHECKER

Every homeowner wants an attractive lawn, healthy shrubs, and towering trees . . . a blended combination of landscaping that will make his home more pleasing, more comfortable, and certainly more valuable.

While fertilizers, weed killers and plant vitamins have their place in gardening, the chief ingredients in landscaping success are good soil and ample water. The purpose of watering is to make up for soil moisture losses. According to lawn authorities, the soil should be moist to a depth of 5 to 6 inches because that is where the permanent grass roots are. Grass absorbs water only through the root ends. If there is no moisture down below, roots will turn upwards to find it. Weeds and annual grasses are encouraged by light sprinklings, but basic lawn grasses are only weakened by it. The amount and frequency of watering depends on the type of soil and the amount of rainfall. Even though the top 2 or 3 inches of soil may be dry, it is still possible that it is sufficiently moist at root level to postpone the next watering for a day or so. On the other hand, flooding the soil is undesirable because it compacts the soil and invites disease.

Obviously, there is more to understand about watering a lawn than turning on the sprinkler . . . and then taking off for a snooze in the hammock! Knowing how much moisture lies at a given depth in the soil can be a great asset in planning lawn care. With the soil moisture checker (Figs. 2-1 and 2-2), you'll get a direct, easy-to-interpret meter reading to help you understand what's happening at root level and to tell you when, where, and how much watering is needed to keep plant life green and flourishing.

The soil moisture checker is a simple electronic instrument designed to measure electrical resistance and thus help you determine how to keep the moisture level of the soil at a desired point. Dry earth is a very poor conductor of electricity. In fact, it is almost an insulator. Add water in increasing amounts, and its resistance decreases, becoming almost negligible when the soil is saturated. Certain types of soil, under extremely moist conditions, might even

Fig. 2-1. Checking the soil moisture.

Fig. 2-2. The Soil Moisture Checker.

become good conductors. In between these two extremes, soil resistance varies over wide limits—in direct porportion to the quantity of water trapped within it. These characteristics are essentially the same for silt, sand, clay, and loams. Hence, by measuring soil resistance, it's possible to know how moist the soil is.

The homeowner will find that there are three basic kinds of soil:

1. Sandy soil—loses water quickly, absorbs heat easily, contains a lot of air.
2. Clay soil—contains very fine particles; is sticky, heavy, and dense; lacks air; holds too much water; becomes muddy after a rainfall.
3. Loam—An intermediary blend between sandy and clay soil, containing humus and lime. Most satisfactory for general use.

Soil resistance is determined by monitoring the minute flow of current between two electrodes (probes) which are inserted into the soil. The soil moisture checker can also be used with indoor plants where insufficient or improper drainage frequently causes a "flood" condition at root level, while dry top soil seems to indicate an obvious need for water. Before too long, such an overwatered plant will surely die.

Construction Details

The soil moisture checker is housed in a standard card-file case (Fig. 2-3) that is available in metal or plastic at any dime store or office supply house. This type of case measures about 5¼ inches × 3¼ inches × 3½ inches overall.

The meter, and the other components it supports, is mounted on a small panel made from a 5-inch × 3⅛-inch piece of plastic, plywood, or hardboard (Fig. 2-4). This panel is supported by a pair of wooden rails cut from two scraps of wood. The rails are 3⅛ inches

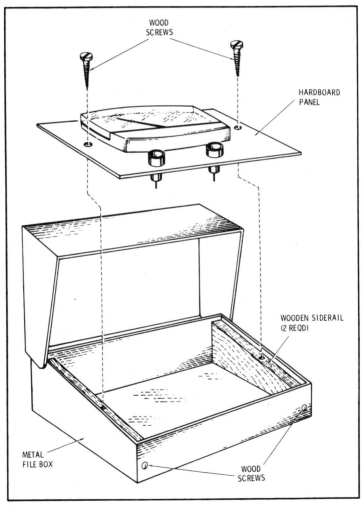

Fig. 2-3. Housing for the Soil Moisture Checker.

Fig. 2-4. Meter and components mounted on hardboard panel.

long and ⅝ inch thick, although thicker stock can be used. The height decreases from 2¼ inches to 2 inches to follow the taper of the sides of the case. The rails are held in the case by screws. Drill a hole at the center of each side of the panel and install a wood screw, tapping it into a side rail on each side. Check the fit of the complete chassis by sliding it into the card file box. Trim and sand as necessary until a comfortable fit is achieved. Now, drill two holes to accommodate the meter mounting screws. The drilling dimensions are shown in the accompanying illustration of the meter. A 1 13/16-inch circular cutout is required to accommodate the body of the meter.

Before installing the meter on the panel, drill two ¼-inch holes near the bottom lip of the panel, to accept the tip jacks. Slip the meter mounting screws through the holes in the panel and gently push the meter down until it seats squarely against the panel. Tighten both nuts to a snug fit.

Trim the leads of resistor R1 to a length of ¾ inch on either side and connect it between the left-hand terminal lug of the meter (positive terminal) and the tip jack directly below, as shown in Fig. 2-4. Solder the lead to the tip jack but don't solder the lead at the meter terminal lug yet. Resistor R2 connects between the left-hand terminal lug and the right-hand terminal lug. Cut its leads to a length of ½ inch on either side, and wrap securely about each lug. Attach the black (negative) lead from the battery clip to the right-hand terminal lug. Now, solder both the left and right terminal lug connections. Complete the wiring by connecting the red (positive) lead from the battery clip to the right-hand tip jack, as seen from the rear of the meter. The schematic diagram for the moisture checker is shown in Fig. 2-5.

Check the wiring carefully. Place the battery next to the body of the meter and slip an elastic band or a length of cellophane tape around its body, so that the band (or tape) falls between the two terminals atop it. This band will hold the battery firmly in place. Snap the battery clip onto the battery. Slip the assembled panel into the case and secure it with two wood screws to the two side rails. Construction of the soil moisture checker is now completed.

Meter readings corresponding to soil conditions are shown in Table 2-1. The probes that are employed with the soil moisture checker are ordinary electronic voltmeter probes. The probes can be pushed into the soil to a depth of several inches.

If you wish to take moisture readings from deeper into the soil, extensions for these probes can be made by using ordinary pointed

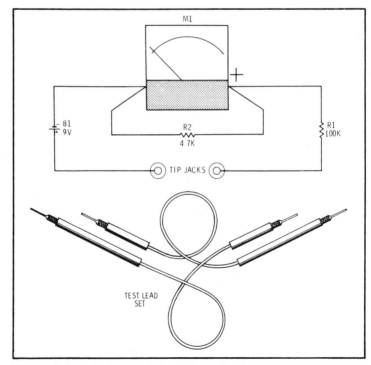

Fig. 2-5. Schematic diagram of the Moisture Checker.

Meter Reading	Soil Condition
10 or less	arid
10 to 18	dry
18 to 26	slightly moist
26 to 36	very moist
36 to 48	wet
48 to full scale	saturated

coat hanger wire, soldered to the tip jacks. To make such a probe extender, cut an 8-inch length of painted wire from a wire coat hanger and remove ½ inch of the protective paint from each end. Bend one end of each wire into a right angle and pass it through the tip of the terminal of each jack. Apply solder so that a sound connection is formed between the tip jack terminal and the extender. These simple extensions can be used for probing deep into the soil. They are strong and reliable and guarantee that soil contact will be made only at root level.

To test the battery in the soil moisture checker, touch the probes together. The meter pointer should indicate no less than 50. If the indication is less than 50, the battery is weak and should be replaced.

A list of parts needed for the soil moisture checker is given in Table 2-2.

Table 2-2. Soil Moisture Checker Parts List

B1—9-volt battery
M1—50-μA meter (Calectro DI-910, Radio Shack
 270-1751, or equiv)
R1—100,000-ohm, ½-W, ±5% resistor
R2—4.7k-ohm, ½-W, ±5% resistor
Case—metal or plastic index card file 5¼″ × 3¼″ × 3½″
Misc—plastic, wood, or hardboard panel 5″ × 3⅛″; side
 rails; wood screws; elastic band; battery clip for 9-volt battery

3. TELEPHONE AMPLIFIER

Want the family to listen in on calls from distant relatives? This simple telephone amplifier (Fig. 3-1) plays both sides of a phone conversation clearly, with no need for cumbersome inductive pickups or acoustic couplers.

This unit does not need to be near the actual telephone as is

Fig. 3-1. The Telephone Amplifier.

required when other coupling methods are used. All you need is an extension jack (as provided by the phone company) and everyone can listen! The volume control selects a comfortable listening level, and a built-in jack even provides for the use of an additional, external extension speaker. Thus, the amplifier can be in one room and an extra speaker in the other, so that both sides of a phone conversation can be heard in a number of different locations. The amplifier is so designed that the volume from the main unit is not affected when the extension speaker is used.

The unit just plugs into your phone line with a coupling capacitor, a matching transformer, and a disconnect ganged with the power switch, all ensuring that this amplifier won't disturb your telephone line. The only precaution: Don't operate it where a phone is close enough to pick up its output—if you do, you'll get a feedback howl.

How It Works

Refer to Fig. 3-2. Two 1-microfarad capacitors in series with coupling transformer T2 connect to the telephone line. These capacitors must be wired with the indicated polarity for proper

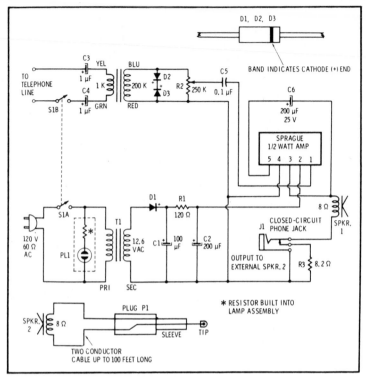

Fig. 3-2. Schematic diagram of Telephone Amplifier.

operation. The transformer provides a step-up of the phone line signal, so that even on long distance calls you'll get enough signal to be heard clearly through the loudspeaker. Diodes D2 and D3 across the transformer's output winding protect the amplifier's input, in case high-voltage surges appear across the line (as they will if the amplifier is connected while the phone is ringing). The amplifier itself is a ready-made, inexpensive Sprague module, which requires no output transformer but only a coupling capacitor between it and the speaker. As the photo shows, the terminals are clearly marked to simplify wiring.

Power is provided by a miniature 12-volt filament transformer, T1, with its output rectified by diode D1 and filtered by C1, R1, and C2. The power switch, S1, is a double-pole, single-throw (dpst) type; one section controls the ac power, and the other connects the unit to the telephone line. When the amplifier is not in use, switching the power off automatically disconnects the amplifier from the telephone line and reduces the load on the phone line, so that normal

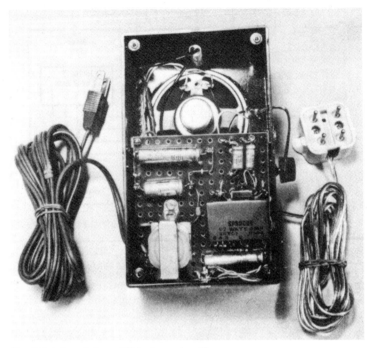

Fig. 3-3. Internal view of Telephone Amplifier.

Table 3-1. Telephone Amplifier Parts List

C1—100-μF, 25-V electrolytic capacitor
C2, C6—200-μF, 25-V electrolytic capacitor
C3, C4—1-μF, 50-V electrolytic capacitor
C5—0.1-μF, 200-V Mylar capacitor
R1—120-ohm, $\frac{1}{2}$-W, ±10% resistor
R2—250k miniature potentiometer
R3—8.2-ohm, $\frac{1}{2}$-W, ±10% resistor
T1—120-V to 12.6-V, 300-mA power transformer
T2—audio transformer (1k:200k)
AMP—Sprague 207C1 $\frac{1}{2}$-W amplifier
D1, D2, D3—general-purpose, 1-A, 50-piv diode
TP—telephone plug
PL1—neon pilot light assembly
S1—dpst toggle switch
P1, J1—matching ultraminiature plug and jack
SPKR—2$\frac{1}{2}$" permanent magnet speaker
Cabinet—3$\frac{3}{4}$" × 6$\frac{1}{4}$" × 2" plastic box (Radio Shack
   270-627)
Misc—line cord, knob, perf board

phone operation is not affected when the amplifier is not in actual use.

Construction

The amplifier is snugly housed in a 3¾-inch × 6¼-inch × 2-inch plastic box, with components mounted on a 3½-inch-square piece of perf board angled within the case. The speaker may be glued or screwed in place and should be mounted first, since the speaker baffle holes in the box determine its location.

The switch, pilot light, volume control, and optional extension speaker jack may be mounted wherever convenient; the locations shown in the photos proved convenient for the component layout shown, though you might wish to mount the pilot light closer to the on-off switch, and the volume control on the front panel near the switch. Notch the perf board to fit around the box's corner posts and locate positions for the "L" brackets that hold the board before you mount the components. Connection to the telephone line may be made through a standard telephone plug, or through a combination male/female telephone jack plug, such as the one listed.

The case we used comes with no back, so we made one from perf board, gluing a thin piece of foam rubber to the underside to cover the screwheads and protect the surface of the desk our amplifier sits on. If foam isn't convenient, "moleskin" foot dressing will do.

4. FLOOD ALARM AND LIQUID LEVEL OR TEMPERATURE MONITOR

Here's a versatile and useful project that can be used to do a variety of jobs. Its possible functions are limited only by the imagination of the builder, and it can be built for a total parts cost of around $15.

The monitor (Fig. 4-1) can be used to sense—remotely if you wish—the level of any liquid in any container. The monitor's probe is normally immersed in the liquid. As soon as the liquid level drops below the probe's tip (where the thermistor is located), a light goes on or a bell or buzzer sounds. If you are interested in meteorology, the monitor can tell you when inches of rainfall have reached a certain preselected level. The monitor can also be used to alert you to the fact that a fluid level (in a fish tank, for example) has dropped below a set level because of evaporation or a leaky container. It can even warn you when potted house plants or greenhouse systems need watering by signaling when the moisture level has dropped

Fig. 4-1. The Flood Alarm and Liquid Level or Temperature Monitor.

below the plant's root system (you simply bury the thermistor probe at the root level and plug in the device). Similarly, it will indicate moisture content in any porous material and, with proper calibration, will even detect and monitor a vacuum by sensing the difference in thermal conductivity between air and lack of it. You merely adjust the self-heat factor of the probe to do this job. As a heat probe, it can also be used as a defrost warning in a freezer. The monitor can then be set to go on when a previously selected (higher than freezing) temperature is reached. In fact, there seems to be no end to the tasks that this little monitor can perform for you.

The actual sensing is done by a tiny passive element about the size of a pinhead, called a *thermistor.* This term is made up from the two words *thermo* and *resistor* because of the thermistor's unique relationship between temperature and its resistance. Thermistors fall into a special class of semiconductors that exhibit a rapid, but extremely large, change in resistance to relatively small changes in temperature.

The basic thermal characteristics of semiconductors were discovered and investigated more than a hundred years ago. As early as 1834, Michael Faraday noted and recorded the high "negative temperature coefficient of resistance" of certain semiconductors. His discovery was nothing more than curiosity at the time, and it wasn't until World War II that industry found uses with sufficient production requirement for these curious devices. Since then, many applications for thermistors have been developed.

It was discovered that a thermistor's characteristic behavior is based on the fact that the *temperature* of a piece of intrinsic semiconductor material determines the number of electrons and "holes" (spaces left in the structure by departed electrons) available for electric conduction. Therefore, the resistance of such a material depends on temperature.

Thermistors are formed under pressure from very pure oxides of nickel, manganese, iron, cobalt, copper, magnesium, titanium, and other metals sintered in a special atmosphere of carefully controlled temperature. The combinations of these materials determine the characteristic of the thermistor. The sintered pieces are then silvered on each side and leads are attached. Thus, a thermistor is a completely passive device consisting simply of two leads joined by a piece of suitable semiconductor material. This results in a component that has a high temperature coefficient of resistance which is nonlinear and negative. In our case, the thermistor is heated by a direct current that is passed through it from the monitor's power supply. After a brief stabilization period, the thermistor temperature reaches equilibrium, *based on its surrounding medium.* It is able to heat up to a higher temperature in air, than in liquid, because the thermal conductivity of air is far poorer than that of a denser liquid; hence, the "self-heat" of the thermistor is contained in its body when in air, but is conducted away when it is immersed in a liquid of moderate temperature.

How It Works

Refer to Fig. 4-2. The monitor's power is applied through a transformer, T1. Its secondary provides us with 12 volts ac which is rectified by diode D1 and filtered by capacitor C1 in conjunction with R5. This gives us approximately 15 volts dc, which is applied to a series circuit consisting of thermistor R2 and the parallel combination of resistors R1 and R3. Transistor Q1 acts as a switch whose state is determined by the current flow into its base-emitter junction. That current is determined by the setting of potentiometer R3, which is first set so that just enough current flows into the base to cause the transistor to switch on (conduct) when the thermistor is in contact with air.

The transistor collector is in series with the coil of sensitive reed relay K1. When the resistance of the thermistor goes down, the voltage at the base of Q1 rises. When the base current reaches the preset level, the transistor suddenly conducts and passes current through the reed relay coil, closing the reed relay contacts by means of the electromagnetic field that surrounds the coil. The current at the base of transistor Q1 is determined primarily by the

18

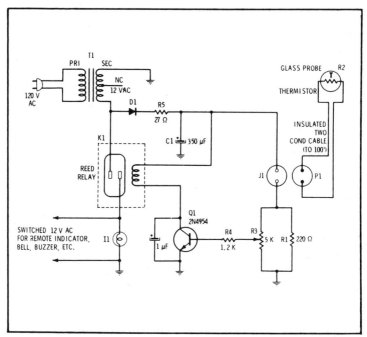

Fig. 4-2. Schematic diagram of the Flood Alarm or Temperature Monitor.

resistance of the thermistor which is, in turn, decided by the environment into which the thermistor is inserted.

Construction

The unit is completely self-contained and housed in a 4¼-inch × 3-inch × 2⅛-inch aluminum case (Fig. 4-3). The transformer is held to the U-shaped channel with two mounting screws. The perf board on which all of the components are mounted is held to the same surface by two 1-inch standoff terminals. The line cord exits through the rear part of the case, and a 12-volt pilot light is mounted on the front. On the same front surface is jack J1 which is a two-terminal jack. The thermistor is mounted in a thin plastic case from a discarded marker pen and is held in place with epoxy cement. When the monitor is not in use, the sensitive thermistor can be protected by the pen's cap.

Calibration

The monitor is calibrated in just a few minutes. Connect the

Fig. 4-3. Unit mounted in a case.

thermistor probe to the monitor and plug the line cord into a 120-volt ac line. Allow about one minute for the circuit to stabilize with the thermistor in air. Now adjust R3 to the point where the pilot

**Table 4-1. Flood Alarm and Liquid
Level or Temperature Monitor Parts List**

D1—1N5062 silicon diode (Radio Shack
   276-1102 or equiv.)
C1—350-μF, 15-Vdc electrolytic capacitor
C2—1-μF, 50-Vdc electrolytic capacitor
K1—spst dry reed relay (Magnecraft W 101MX-4)
Q1—2N4954 npn silicon transistor (Radio Shack
   276-2009, or equiv)
R1—220-ohm, ½-W, ±10% resistor
R2—glass probe thermistor, 1000 ohms @ 25°C (Fenwal
   Electronics Type GB 31P2)
R3—5k miniature potentiometer (Mallory, MTC53L1 or equiv)
R4—1.2k, ½-W, ±10% resistor
R5—27-ohm, 1-W, ±10% resistor
T1—12.6-V, 2-A filament transformer (Stancor P-8130, Radio
   Shack 273-1505, or equiv)
I1—12-V indicator lamp
P1—Amphenol plug Type 80MC2M
J1—Amphenol jack Type 80MC2F
Case—aluminum minicase 5¼″ × 3″ × 2⅛″ (Premier Metal
   Products Corp. Type PMC-1066, Radio Shack 270-238, or equiv)
Misc—perforated phenolic board, hardware, wire, solder etc.

light just goes on. Insert the probe into a liquid (water will be just fine) and the lamp will go out after a few seconds. If it does not, adjust R3 so that the lamp just goes out. Then take the probe out of the liquid and note that after about seven seconds or so, the lamp goes on, indicating that the probe is in air. It may be necessary to go over this adjustment once or twice, but once set, it will hold.

To use the monitor in a refrigerator or freezer, affix the probe to a 1-inch × 1-inch piece of metal and place this combination inside the compartment you wish to monitor. After a stabilization period, adjust R3 so that the light goes on with rising temperature but remains off when the temperature is as low as required.

A parts list is given in Table 4-1.

5. TARRY LIGHT

Here's an ac light switch we're willing to bet is different from any you've ever seen. The reason—it sports not one, but two, control switches (Fig. 5-1). The first—the illuminated seesaw switch—works like any conventional light switch. Flip it one way and you turn the light on; flip it the other way and you turn the light off. The second switch (the push button on top) . . . well, that's something special, and so is the potentiometer. Together, the push button and potentiometer initiate a time delay that turns a light on (just touch the button), then automatically turns it off again after a predetermined time. How long? That's up to you. The potentiometer can be set for a delay of a few seconds, to just under three minutes.

In any house, and probably yours too, there are several lighting control situations where the time-delay feature is worth a pound of shoe leather in saved steps. For example:

1. In a long corridor not equipped with two-way switches.
2. In a garage. With the Tarry Light, there's no need to walk back through the garage to reach the light switch on the rear wall when taking the car out at night.
3. To control a front (or driveway) floodlight. The Tarry Light gives you plenty of time to find your way through the front yard (or down the driveway) before the floodlight turns off automatically.

We've even made it easy to find the Tarry Light in the dark. When the unit is off, the built-in neon lamp in the seesaw switch is

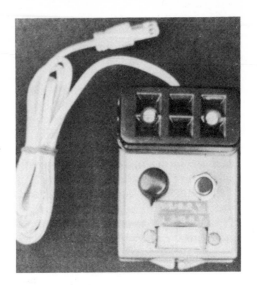

Fig. 5-1. The Tarry Light.

lit. No fooling and no fumbling—the Tarry Light makes it easy for you to solve your lighting control problems.

How It Works

Refer to Fig. 5-2. The time-delay circuit is built around a C106B1, a 200-V, 2.5-A (rms) silicon controlled rectifier. Here's what happens: Whenever you press the push-button switch SW2, capacitor C1 gets charged through D5 to the full dc voltage developed by the diode bridge comprising D1, D2, D3, and D4. Release the button and the charged capacitor is now connected across the series combination of R2, R3, and potentiometer R4 whose setting determines the total resistance and thereby sets the time it takes for the capacitor to discharge. A steering diode, D6, connected to the junction of R2 and R3, picks off a portion of this decaying dc voltage and applies it to the gate terminal of Q1, the SCR, triggering it into a conductive state. This SCR will remain *on* as long as there is sufficient voltage on its gate. As soon as this voltage decays below the minimum holding voltage of the SCR, it will turn *off* on the next line alternation.

The SCR acts as a switch across the diode bridge; so when it is *on*, it effectively connects points A and B so that any load plugged into J1 will receive full 120 Vac, but only for *as long as the SCR is conducting.* The load current flows through D2 and D4 during one-half of the cycle and through D1 and D3 during the other

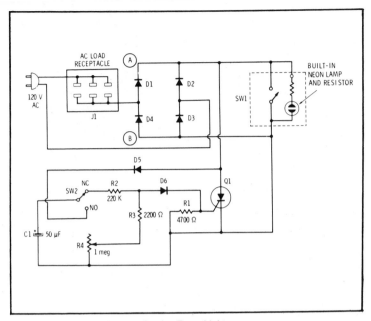

Fig. 5-2. Schematic diagram for the Tarry Light.

half-cycle. Switch SW1 serves as an ordinary on-off switch. We selected an illuminated switch for ease of use. Any load not greater than 250 watts—as limited by the SCR—can be powered by the Tarry Light. It is clear from the schematic that a load must be plugged in to test the unit.

Construction

The Tarry Light is housed in a small plastic case measuring 1⅝ inches × 3⅝ inches × 1⅜ inches (Fig. 5-3). The illuminated seesaw switch mounts through a rectangular cutout. Push-button switch SW2 and potentiometer R4 mount right above it. A small ac outlet, into which three separate loads can be plugged, mounts on the remaining space on the bottom of the case which serves as the top surface of the Tarry Light. Connections to the outlet are made with wires that pass through two holes in the case. The line cord exits through a small hole in the rear. The four diodes, D1 to D4, mount on a small terminal strip that also serves as a support for the anode of the SCR. Figure 5-3 shows clearly how the remaining parts are wired into the circuit. The parts list for the Tarry Light is given in Table 5-1.

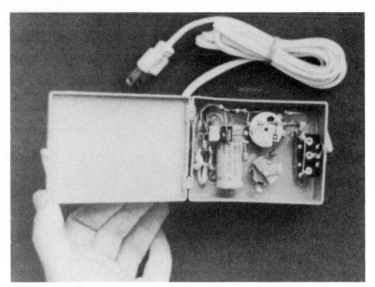

Fig. 5-3. Housing for the Tarry Light.

Table 5-1. Tarry Light Parts List

Q1—SCR, GE Type C106B1 or Radio Shack 276-1067, or equiv
D1 through D6—3-A, 200-V general-purpose rectifier diode
R1—4700-ohm, ½-W, ±10% resistor
R2—220k, ½-W, ±10% resistor
R3—2200-ohm, ½-W, ±10% resistor
R4—1-megohm potentiometer (Mallory Type U54 Midgetrol or equiv)
C1—50-μF, 150-Vdc electrolytic capacitor
J1—ac outlet
SW1—illuminated seesaw switch and built-in neon lamp and resistor (Radio Shack 275-670)
SW2—spdt normally open push-button switch (Radio Shack 275-1549)
Misc—plastic case 1⅝″ × 3⅝″ × 1⅜″, line cord, knob

6. TELL-A-BELL

Have you ever missed an important telephone call because you were working outdoors, or in the garage, or lounging on the patio or porch, where you couldn't hear the bell? Most of us have had this frustrating experience. One way to avoid it is to have the phone company install a loud outside bell. This is fine if you need the outside bell all year round, and if it won't disturb the neighbors, and if you're always in an area where the bell can be heard readily. But that's a lot of "ifs," which don't seem to justify the expensive installation by the phone company of a loud outside bell.

Here's where the Tell-A-Bell (Fig. 6-1) comes to the rescue! This handy accessory safely connects directly to your phone line by means of a standard phone jack and plug, and will activate an attention-getting 120-V bell, buzzer, or horn whenever your phone rings. Use it when you need it; disconnect it when you don't. If you prefer to "see" rather than hear the ring, you can use the Tell-A-Bell to make a light (such as a table lamp) go on and off until the phone is answered. This form of signaling is of particular value to the hearing-impaired, and also to anyone who wants to turn off the phone's bell so as not to wake a baby, or disturb a light sleeper or a sick person, yet who doesn't want to miss a phone call. All you have to do is plug a light into the Tell-A-Bell output and turn off the telephone bell. On incoming calls, it will make the light go on and off until you answer. With this mode of operation, however, you must keep the activated light source within sight.

How It Works

Refer to Figs. 6-2 and 6-3. The Tell-A-Bell employs a unique, optically coupled, electrically isolated transducer to activate the

Fig. 6-1. The Tell-A-Bell.

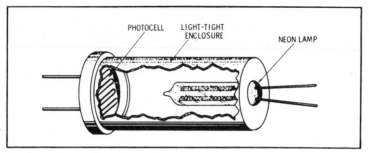

Fig. 6-2. Light-tight enclosure.

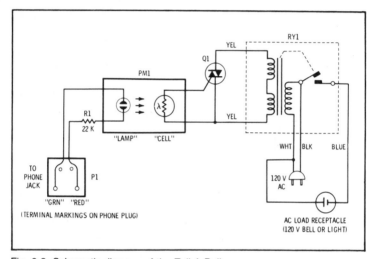

Fig. 6-3. Schematic diagram of the Tell-A-Bell.

circuit when a ringing voltage appears on the phone line. This four-terminal transducer, called a Photomod®, comprises a neon bulb in close proximity to a photocell. Both of these are enclosed in a light-tight tube, as shown in Fig. 6-2. When the lamp is off, the cell is dark and its resistance is very high. However, when a ringing voltage appears on the phone line, the neon bulb glows brightly and illuminates the photocell whose resistance then drops to around 1000 ohms.

The series arrangement of R1 and the extinguished neon bulb in the Photomod are essentially an "open circuit" across your phone line. Hence, the Tell-A-Bell can be left on the line at all times. It not only will be "invisible" on the phone line when there's no conversation going on, but it will also not load down the line when the phone is actually in use.

26

The photocell is placed in the gate circuit of a triac so that it will turn on whenever the cell resistance drops. The triac is connected directly across the switching terminals of the isolation relay. Thus, the relay closes and applies 120 volts to the output socket whenever the triac is on, effectively shorting the switching winding.

Together, the Photomod and the isolation relay give *double* protection to the phone line so that there is no chance of a line voltage being fed back into the telephone circuit. This is a most important safety feature of the Tell-A-Bell.

The circuit is capable of supplying a maximum of 7 amperes at 120 Vac to any load plugged into the ac outlet of the Tell-A-Bell.

Construction

The transformer and the few associated electronic components fit neatly into a small phenolic box measuring 3¾ inches × 2⅝ inches × 1⅜ inches (Fig. 6-4). The triac, the Photomod, and the 22,000-ohm current-limiting resistor for the neon bulb mount easily on a 1⅛-inch × 2⅜-inch perf board which is held in place by a screw/spacer arrangement through the bottom of the box. The isolation relay is held in place by two screws through the bottom of the case. You must use the parts specified to ensure a safe installation that won't be hazardous to you or phone company personnel.

Connection to the Phone Line

If you already have a phone jack installed by the telephone company, simply wire a plug with through-terminals so that you can

Fig. 6-4. Components mounted in case.

plug in the Tell-A-Bell; then plug the phone into the Tell-A-Bell's plug, totem-pole fashion. When you don't want the Tell-A-Bell function, just unplug it and plug in your phone for normal use.

A parts list for Tell-A-Bell is given in Table 6-1.

Table 6-1. Tell-A-Bell Parts List

Q1—triac RCA 40526
R1—22k, ½-W, ±20% resistor
RY1—isolation relay ALCO FR 105 (do not substitute)
PM1—Photomod Clairex CLM3120A
P1—phone plug
Case—phenolic 3¾″ × 2⅝″ × 1⅜″ (Calectro J4-725, or equiv)
Misc—ac outlet, wire, line cord

7. GARAGE STOPLIGHT

About the last thing you'd like to run into in your garage is the rear wall, or junior's bicycle, or the patio furniture you've stacked for the winter. But, with household storage space at a premium, things may be moving into your garage so quickly that easing your car into place requires the nimble reflexes of a pickpocket. Go too far and there's trouble—not far enough and you won't be able to close the garage door.

Here's a unique gadget that can help. It's a garage stoplight (Fig. 7-1) that senses when your car has reached just the right spot. It signals you by flashing a light, then resets for the next time. The stoplight is actuated when your car's front tire comes to rest on a narrow strip of tape that acts as a sensing switch (Fig. 7-2). This sensing switch, called a ribbon switch, contains a normally open spst switch that is *2 feet long*, and it takes only 12 ounces of pressure to close it. The ribbon switch is cemented to the garage floor as

Fig. 7-1. The Garage Stoplight.

shown in Fig. 7-2 so that when the front tire first rolls over it, the car is *exactly* where you want it to stop. Once you've got the right spot for the switch, parking becomes a cinch; hit the switch, and the light tells you to STOP!

How It Works

Refer to Fig. 7-3. Capacitor C1 is permanently connected across the 3-volt supply (B1 and B2 in series) through 10-megohm resistor R1. This resistor allows the capacitor to charge (relatively slowly and with very little drain from the two dry cells) to 3 volts. The instant switch SW1 is closed, it connects the charged capacitor (C1) in series with C2 and R2. Capacitor C2 starts to charge, placing

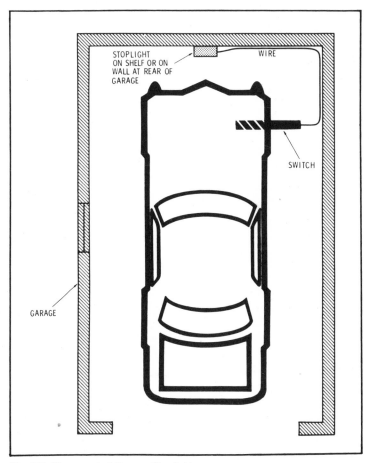

Fig. 7-2. Placement of Garage Stoplight.

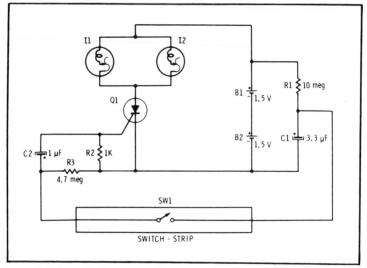

Fig. 7-3. Schematic diagram for Garage Stoplight.

a positive-going voltage on the gate of the SCR and causing it to turn on. As soon as this happens, the two parallel-connected bulbs I1 and I2 turn on. These little bulbs are of a unique variety. They are called "self-flashing" bulbs because they contain a tiny strip of bimetal in series with the filament, which tends to bend slightly when heated—as is the case when current flows through this little strip and the filament. Because of the internal construction of the bulb, this slight bend interrupts a connection within the bulb and it goes out. When the strip cools, it goes back into place and the bulb is once again ready to go on. Because the bulbs are connected in parallel and are not *precisely* identical in their performance characteristics, one of them will go out before the other. It will go on again, then the other bulb will alternate until such time as they both happen to be *off* at the same time. In the meanwhile, capacitor C2 has become fully charged (if SW1 remains closed, for example), and since there is no more current through R2, there is zero voltage on the gate of the SCR. This would also be true if S1 were closed only momentarily.

It is a well-known fact that an SCR in a dc circuit will remain in an on (latched) state once it is triggered unless it is "commutated." Then it will go off. In our case, the SCR is commutated by the self-flashing bulbs. Once they break the circuit, the SCR stops conducting and won't go on again until a pulse appears once more on its gate. This method of commutating an SCR is covered by US letters Patent 3,681,753.

Construction

The two warning lights, as well as the handful of other components, are all easily fitted inside the housing (Fig. 7-4). The two dry cells that power the stoplight are held in a standard plastic battery holder which is affixed to the backside of the housing with a dab of RTV or other suitable cement (Fig. 7-5). The battery holder is placed so that the stoplight will stand steady when placed vertically on a shelf in the garage.

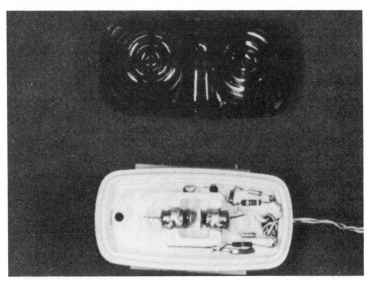

Fig. 7-4. Lights and components fit in the lens housing.

Fig. 7-5. Batteries are housed in plastic battery holder.

The marker light comes with one lead which connects to the positive terminal of the battery holder. A wire from the negative terminal feeds through an eyeletted hole to the junction point of C1, R2, and R3 which is self-supporting. The bulbs are soldered in place. They are automatically connected in parallel because of the way the marker light is constructed. The other small components as well as the SCR are easily fitted in place as shown by the photograph in Fig. 7-4. The two wires from the stoplight to the floor switch pass through the same hole as the wire to the battery holder.

A parts list for the Garage Stoplight is given in Table 7-1.

Table 7-1. Garage Stoplight Parts List

R1—10 megohm, ½-W, ±10% resistor
R2—1k, ½-W, ±10% resistor
R3—4.7 megohm, ½-W, ±10% resistor
C1—3.3-μF, 10-V tantalum capacitor
C2—1.0-μF, 10-V tantalum capacitor
SCR—GE Type C103 silicon controlled rectifier or Radio Shack
    276-2009
I1, I2—self-flashing bulb, No. 401 (1.25-V/0.22-A)
SW1—Tapeswitch ribbon 2 feet long
B1, B2—size C, 1.5-V cells (alkaline)
Marker light—Signalstat 1211
Battery holder—Calectro F3-063

8. THREE-MINUTE TIMER

Is talk cheap? Just take a look at your phone bill. As time flies, Ma Bell bites deeper into your purse, providing the inescapable truth: Time *is* money.

The reward for concise conversation on the telephone is measurable in dollars and cents. And, in this day when our government has learned how to cut the dollar in half without touching the paper, we could *all* take pleasure in a lower phone bill!

Here is the answer to compulsive and costly chatter: an electronic timer that sounds off *exactly* three minutes after it is activated (Fig. 8-1). You turn it on at the beginning of a conversation by setting its switch to the time position. Three minutes later, its gentle but persistent "bee-e-e-e-p" reminds you that the time is up. You can turn it off, or you can reset it and be reminded again after another three minutes have elapsed, by flicking its switch to the off position and then back to the time position. The unique beeping sound is produced by a Sonalert®, which is a self-contained transistorized sound source.

Completely portable and battery-operated, the timer has other uses besides curtailing your teenage daughter's telephonic tête-á-tête. In the kitchen it guarantees a perfect 3-minute egg. For chess and word-building games (like Scrabble or Spill-N-Spell), use it to

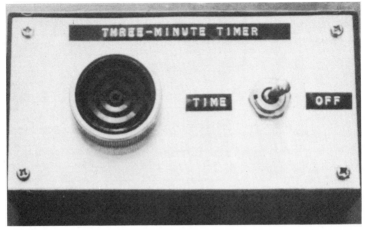

Fig. 8-1. The Three-Minute Timer.

time each player's turn. Check your reading speed by scanning a page while the timer counts down seconds. Count the words read, divide them by three, and you'll have your *average* reading rate in words-per-minute.

How It Works

Refer to Fig. 8-2. When S1 is in the off position, capacitor C1 charges to within half a volt of the battery voltage through diode D1 and resistor R4. This means that the voltage developed across C1 appears across a resistive ladder comprising R1, R2, R3, and R4.

When S1 is closed (in the time position), the anode of the programmable unijunction transistor (PUT) rises to the positive supply voltage. The PUT does not conduct, however, because the battery voltage now appears in series with the charge stored on C1, raising the gate of the PUT to a level quite positive with respect to the anode. In order to conduct, the PUT gate must be at a voltage that is negative with respect to the anode voltage. (In other words, if the anode is at +9 volts, the PUT will reliably trigger if the gate is at +8.5 volts.)

Thus, the timer relies on the *discharge* of capacitor C1 through resistors R1, R2, R3, and R4. For, once C1 is at zero volts, the battery voltage (less the forward drop across D1) will guarantee that the gate voltage is negative with respect to the anode. At this point, the PUT will turn on and "latch" (like an SCR), applying battery voltage to the Sonalert and causing it to sound.

The resistor values of R1, R2, and R4 are chosen to bring the

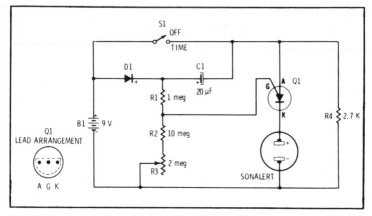

Fig. 8-2. Schematic diagram of the Three-Minute Timer.

discharge time of C1 into the three-minute range. Resistor R3 is adjusted for the value needed to obtain a precise three-minute time constant.

Once the Sonalert sounds, opening S1 will break the PUT latch by interrupting the path for battery current flow. Meanwhile, C1 will recharge instantly through D1 and R1. Closing the switch, again, commences a new time cycle.

Construction

Construction is simple, and you won't need a fully equipped machine shop. The small, plastic utility case comes with an aluminum cover plate, and only two holes need be drilled. The Sonalert mounts in a 1⅛-inch-diameter hole; the toggle switch slips nicely into a 7/16-inch-diameter hole. To make the mounting hole for the Sonalert, scribe a 1⅛-inch-diameter circle on the case, using a compass, and drill a tiny pilot hole at the center of the circle. You can now use a hole saw to cut the required diameter hole. If you prefer, you can use a chassis punch available at low cost from most electronic mail order houses, but unless you plan to do a lot of building, you may not wish to invest in a chassis punch. You can do just as good a job with a simple technique that takes only a bit longer and requires just a little more elbow grease.

Inside the first circle, scribe another circle having a 1-inch diameter. Now, fit a ⅛-inch drill bit into your electric drill and carefully drill a series of holes all around the ⅛-inch-wide track formed by the two concentric circles. As you drill, space the holes as closely together as you can. When you've finished, you'll have a circle of holes separated by thin slivers of metal. Cut through these slivers with a chisel, or an old screwdriver, lightly tapping with a hammer. The metal slug will drop right out, leaving a hole in the aluminum panel almost exactly the right size. Smooth the irregular edges of the hole with a file and you'll have a perfect 1⅛-inch-diameter hole in which to mount the Sonalert.

The hole for the switch can be drilled and reamed or filed to size. Use either a ¼-inch or ⅜-inch drill bit (depending upon the maximum your electric drill can accept) to make the starting hole for the switch. Now use a tapered reamer, or an old triangular file, to enlarge the hole to the required 7/16-inch diameter.

After checking your construction work, the final detail is to install B1. Insulate the battery's foil case with a piece of cellophane tape applied to the entire length of the narrow side of the case. You can hold the battery in place with a dab of glue or cellophane tape, or else you can slip an ordinary elastic band around the battery and the Sonalert. Connect the battery clips to the terminals of the battery and you are ready to calibrate (Fig. 8-3).

Fig. 8-3. Parts mount on cover of case.

How to Calibrate the Timer

Calibration of the completed timer can be accomplished by using the second hand of an accurate clock, wristwatch, or stopwatch. With the second hand at the 12 o'clock position, set toggle switch S1 to the time position and wait out the interval until the Sonalert sounds. Note the elapsed time—it should lie between 140 and 220 seconds. Calibration simply means adjusting potentiometer R3 until a 180-second countdown interval is achieved. When the Sonalert sounds, set toggle switch S1 to the off position. If the elapsed time interval was less than 180 seconds, adjust the cross-slotted shaft of potentiometer R3 with a small screwdriver, turning it in a counterclockwise direction. How far you turn the shaft depends on how much less than 180 seconds your first trial was. If you were only a few seconds off, turn the shaft counterclockwise only slightly. If you were as much as 30 seconds off, you may have to rotate R3's shaft as much as 60° for your second try. After adjusting R3, flip the switch to the time position, and clock another interval until the Sonalert sounds You should now be much closer to the desired three-minute time span. Successively smaller adjustments of R3 will bring you closer and closer until you have finally adjusted the time to a perfect 180-second period. Of course, the same adjustment procedure holds true if your time period is *greater than* 180 seconds, the only difference being that R3's shaft is adjusted in a clockwise direction.

Once you have achieved a 180-second setting, the adjustment is permanent, and no further calibration will be required, even if you change batteries.

A parts list for the Three-Minute Timer is given in Table 8-1.

Table 8-1. Three-Minute Timer Parts List

B1—9-V battery (Mallory Duracell MN1604, or equiv)
D1—general-purpose silicon diode; 1N645, 1N4005, or equiv
(Radio Shack 276-1104)
C1—20-μF, 50-V electrolytic capacitor (Sprague TE-1305,
or equiv)
R1—1-meghom, ½-W, ±5% resistor
R2—10-megohm, ½-W, ±5% resistor
R3—2-megohm screwdriver-adjust potentiometer (Mallory
MTC26L4, or equiv)
R4—2.7k, ½-W, ±5% resistor
Q1—programmable unijunction transistor (GE D13T1)
S1—spst toggle switch
Sonalert—Mallory SC628 or Radio Shack 273-060
Utility case—5″ × 2⅝″ × 1⅝″ Bakelite box (Radio
Shack 270-233)
Misc—wire, solder, elastic band, battery clip

9. MINI SATELLITE SPEAKER

It's only 2½ inches in diameter, but it booms out a surprising amount of sound for its size. And it's colorful, too—red, yellow, blue, green, white—any shade you want. This miniature speaker (Fig. 9-1) is a plastic spray-can top fitted with a tiny permanent-magnet speaker. It's ideal for use with pocket transistor radios and

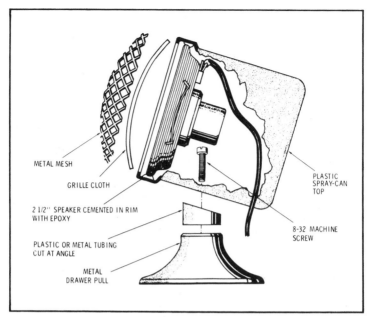

METAL MESH

GRILLE CLOTH

2 1/2'' SPEAKER CEMENTED IN RIM
WITH EPOXY

PLASTIC OR METAL TUBING
CUT AT ANGLE

METAL
DRAWER PULL

PLASTIC
SPRAY-CAN
TOP

8-32 MACHINE
SCREW

Fig. 9-1. The Mini-Speaker.

other small portables that have a jack for attaching an earphone, or as a "slave" station speaker when used with an intercom. When the radio isn't traveling, you can plug in the speaker and keep it on a bookshelf, kitchen counter, or bed table. Though small, the speaker will greatly improve the sound of a transistor radio as well as add a novel appearance.

Plastic caps from household spray cans come in every color of the rainbow. Their inside rim is a perfect fit for a 2½-inch round speaker, cemented in with epoxy or silicone adhesive. A metal drawer pull makes an attractive pedestal. For a slight upward tilt, cut a ring from ¾-inch metal or plastic tubing at a 30° angle to form a sloping collar. We used the tube from a Polaroid print coater and it worked just fine.

Solder two thin wires of the desired length to the speaker terminals. Drill a small hole in the enclosure and pass the wires through it. At the other end of these wires, connect a suitable plug for your radio or intercom.

The grille is a piece of speaker cloth covered with metal mesh. Form the mesh around a tennis ball to give it a little curvature, and cement it and the cloth into the cap's rim over the speaker. Speakers of the type shown are readily available and are very inexpensive.

Build it and you will find that you will have a super sound source.

A list of parts needed is given in Table 9-1.

Table 9-1. Mini Satellite Speaker Parts List

2½″ PM Speaker
Plastic spray can top
Grill cloth, metal mesh, tubing (for tilt)
Metal drawer pull

10. JUNIOR SATELLITE SPEAKER

Want something a little bit larger? Then build the Junior Satellite Speaker (Fig. 10-1). We used a 3-inch PM speaker and a 4½-inch plastic container of the kind used by your favorite deli for potato salad or coleslaw (Fig. 10-2). These containers come in many different colors and are quite attractive. Again, we used a metal drawer pull as the base. The photos clearly show the simple sequence of operations necessary for easy assembly of the speaker.

Fig. 10-1. The Junior Satellite Speaker.

The container is first secured to the drawer pull with a screw and a flat washer. Because these containers are tapered, they will tilt slightly upward when the unit rests on a horizontal surface, and that's just what we want. Cut a small hole in the back for the speaker wires. Next, cut a 2⅞-inch hole in the center of the cover of the container and cement the speaker behind it. This can easily be done with a pair of scissors. Then solder the wires to the speaker

Fig. 10-2. Pedestal and speaker mounting.

terminals. A little epoxy cement around the edges of the speaker will hold it in place. Be careful not to get any of the cement onto the cone of the speaker. This would stiffen it and would take some of the compliance out of the cone, which could result in objectionable audio distortion.

Cut an 8-inch disc from a thin piece of foam rubber (Fig. 10-3).

Fig. 10-3. An 8-inch foam rubber disc for a dust shield.

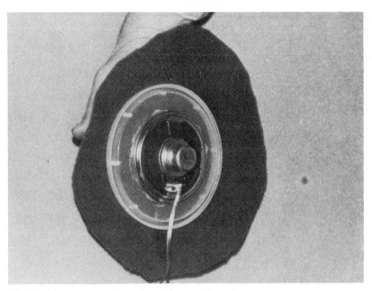

Fig. 10-4. Place cover in center of foam rubber disc.

This will serve as a dust shield and decorative grille, and will also save the cone from being accidentally punctured. Place the cover (with speaker firmly glued in place) in the center of the foam rubber (Fig. 10-4) and start to fold it in all around the edge (Fig. 10-5). Then

Fig. 10-5. Fold foam rubber around edge of cover.

Fig. 10-6. Push speaker and foam rubber into container.

place it in the container, but be sure to have as much of the foam rubber as possible tucked away inside to present a smooth surface on the outside (Fig. 10-6). Next, push the cover in place; this will firmly set the speaker assembly in place and also assure a taut foam rubber covering on the outside.

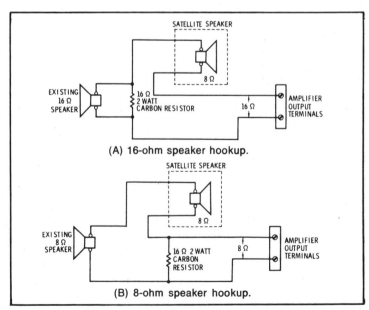

Fig. 10-7. Maintaining proper impedances.

The Junior Satellite Speaker can be used in much the same way as the Mini-Speaker assembly. Either is suitable for use as the slave station speaker of the Squawk-N-Talk intercom project described in Chapter 12.

To maintain the proper impedances in your hookup to an audio system, see Fig. 10-7.

A parts list for the Junior Satellite Speaker is given in Table 10-1.

Table 10-1. Junior Satellite Speaker Parts List

3″ PM Speaker
Tapered plastic container
Thin, round foam rubber disc, 8″ diameter
Metal drawer pull

11. PHOTOCELL MEMORY SWITCH FOR AC POWER CONTROL

Here's a handy little gadget that lets you remotely control any ac-powered device by using the beam of a flashlight as your "magic wand" (Fig. 11-1). Use it to turn your tv set or stereo on or off from the cozy comfort of your bed or easy chair. Or, use it to control room lights, radio, small appliances—anything powered by the ac line—by using a light beam from a handheld flashlight. Or, aim it out a window, and let the morning sun start your electric percolator brewing!

The important part of this gadget is that it "remembers" what you've told it to do. Activate it once to apply power to a device and it stays on. Activate it a second time and power goes off and stays off.

The secret is the combination of a high-sensitivity photocell, a

Fig. 11-1. Photocell memory switch.

high-gain IC Schmitt trigger, and an impulse-actuated latching relay. These components, housed with a simple power supply in a compact plastic case, make an interesting project and a fascinating gadget for which you'll find many uses.

How It Works

The ULN 3304M is an IC Schmitt trigger in a miniature DIP package. This useful little circuit (Fig. 11-2) is a voltage-sensitive switch or, as it is sometimes called, a threshold detector. The IC provides no output (it remains off) until the voltage at the input (pin 7) reaches or exceeds a predetermined level, typically about two-thirds of the supply voltage. When that happens, the circuit turns on and provides an output voltage at pin 3.

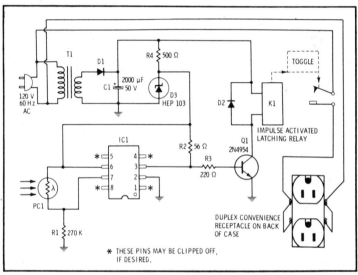

Fig. 11-2. Schematic diagram for Photocell Memory Switch.

Photocell PC1 and resistor R1 form a voltage divider between the supply voltage and ground. The output of this voltage divider is applied to pin 7 of IC1. The cadmium sulfide photocell we use here is essentially a photosensitive resistor, which means that its resistance depends on the amount of light falling on its surface. When there is little or no incident light, cell resistance is very high— greater than 50 megohms in complete darkness of this particular type. When exposed to light, the cell's resistance drops dramatically. The CL602 measures approximately one megohm when il-

43

luminated by a light of 2 footcandles strength, and much less at greater levels of illumination. Thus, when a bright light falls on the photocell, its resistance drops substantially and the voltage on pin 7 of the IC rises to a level where it will turn on the Schmitt trigger. This will now produce an output on pin 3 which is applied to the base of relay driver transistor Q1, so that it too will turn on and thereby activate the relay.

This particular relay functions just like an electrically activated toggle switch. Every time it is activated, it "flips." Thus, on successive impulses, it alternately closes and opens a pair of contacts. Connect this pair of contacts in series with the ac line and you've got yourself a photocell memory switch for ac power control.

The dc voltages required to operate the IC, relay, and transistor are obtained from the rectified (D1) and filtered (C1) output from a small 24-V transformer (T1). Diode D2 across the relay winding protects the output transistor so that it will not be "zapped" by the high voltages that are generated by the collapsing magnetic field around the relay coil whenever it is suddenly turned off.

Resistor R1 determines the light sensitivity of this circuit. A lower value will require more light on the cell to "switch," whereas a higher value may make the unit so sensitive that ambient light will keep the Schmitt trigger on all the time. (Be sure the cover is on the case when you test the unit to determine its sensitivity.)

Construction

The complete unit is housed in a 3¾-inch × 2⅝-inch × 1⅜-inch plastic case. All components, with the exception of the relay and diode D2, fit neatly on a small perf board. Figure 11-3 shows the layout very clearly. The transformer is oriented so that its primary

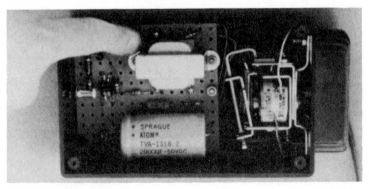

Fig. 11-3. Memory switch mounted in case.

(high-voltage) terminals are on the "outside" of the perf board.

Since we supported the IC by its pins, we clipped off the unused ones—pins 1, 4, 5, and 8. If you should choose to use a socket, then you don't need to clip these pins.

Wiring between components is done on the bottom side of the board, which is held to the bottom of the case with three screws and nuts. You must drill a ¼-inch hole on the front surface of the case to admit light to the photocell. The duplex ac convenience outlet and the relay are affixed to the opposite side of the case. In addition to mounting holes, you will also need two holes for the wires that carry the ac power to the duplex outlet.

Arrange the photocell in such a way that it is behind the ¼-inch opening, but inside the case. In this way it will be affected only by light that is directed deliberately at its surface from your flashlight. Note that resistor R4 does get warm to the touch. That's all right.

A parts list for the Photocell Memory Switch is given in Table 11-1.

Table 11-1. Photocell Memory Switch Parts List

IC—Sprague Electric Type ULN 3304M Schmitt trigger
Q1—2N4954 transistor, or equiv (Radio Shack 276-2009)
R1—270k, ½-W, ±10% resistor
R2—56-ohm, ½-W, ±10% resistor
R3—220-ohm, ½-W, ±10% resistor
R4—500-ohm, ½-W, ±10% resistor
C1—2000-µF, 50-Vdc capacitor (Sprague TVA 131812, or equiv)
D1, D2—general-purpose, 1-A, 200-piv diode
D3—60.2-V, 1-W zener diode (HEP 103, or equiv)
K1—relay, Potter & Brumfield Type PC11D
PC1—photocell, Clairex CL602
S1—duplex outlet
T1—transformer, 120-V/24-V @ 300-mA (Radio Shack 273-1386)
Phenolic case—6¼" × 3¾" × 2" (Calectro J4-723, Radio Shack 270-627, or equiv)
Misc—line cord, 3½" × 3⅝" perf board

12. SQUAWK-N-TALK IC INTERCOM

This single IC intercom (Fig. 12-1) uses just a handful of parts, and its standby current drain is low enough to allow operation from a built-in battery of AA-size alkaline cells. But, when you need mouth-power to communicate over the high noise level of a busy

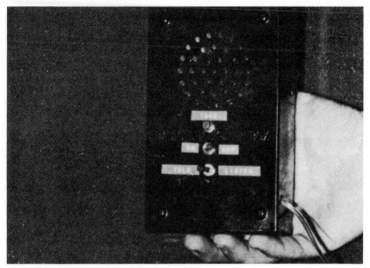

Fig. 12-1. The Squawk-N-Talk IC Intercom.

shop or a house full of kids, this unusual amplifier can deliver peak audio output of up to 2 watts.

The secret is in the IC, a Sprague ULN2280B. It features an internally fixed gain of 34 dB through a complete amplifier consisting of preamplifier and power stages, all on one chip. This power gain is a natural match for the tiny signal available from a permanent magnet speaker used as a microphone. By placing one speaker/mike in a plastic case containing the amplifier (the master station), and placing another speaker/mike (the slave station) at a remote location (up to several hundred feet away), it's possible to talk and listen at just the flick of a switch.

To summon a party to the slave station, the circuit features a "squawk" tone button. Just push the button and the slave speaker will let out a loud tone.

How It Works

Figure 12-2 is a schematic diagram of the Squawk-N-Talk intercom. Speaker SP1 connects through one-half of a dpdt talk-listen switch (S3A) to the secondary of transformer T1, which is used to match the impedance of the speaker to the input impedance of IC1. Speaker SP1 also connects to the opposite section of the talk-listen switch (S3B). The slave speaker, SP2, is similarly, but oppositely, connected to the poles of S3A and S3B. Thus, in the talk position of S3, the master station speaker SP1 acts as a micro-

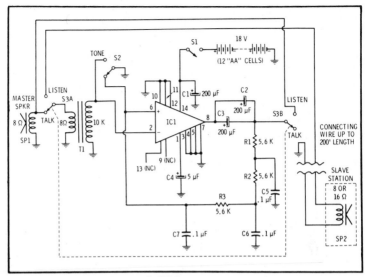

Fig. 12-2. Schematic diagram for IC intercom.

phone, while the slave station speaker, SP2, acts as a loudspeaker. In the listen position, the roles of SP1 and SP2 are reversed.

The 18-volt power supply for the intercom is made up of a dozen 1.5-volt AA-size alkaline cells. These are contained in two 6-cell battery holders which fit easily into the master station's case. The battery supply is decoupled by electrolytic capacitor C1. Decoupling of the internal preamplifier circuitry of IC1 is provided by C4. The amplified audio output of IC1 is coupled through parallel capacitors C2 and C3 to switch S3B, and then to either SP1 or SP2. The 400-μF capacitance of parallel capacitors C2 and C3 provides a good match between the low impedance of the speaker and the amplifier's output. Note that the output of IC1 is also applied to a phase-shifting network consisting of R1-C5, R2-C6, and R3-C7. This network connects back to the inverting input of IC1, which is normally grounded by S2. However, if S2 is opened, the inverting input rises above ground and is connected to the output of IC1 through the network. Amplified noise at the output is now phase-shifted and fed back to the input, causing oscillation at an audio frequency. In effect, the IC audio amplifier becomes a power audio oscillator. If talk-listen switch S3 is in the talk position, a loud tone will issue from the slave speaker, summoning the party you wish to speak to. Releasing S2 grounds the inverting input of IC1, stopping the feedback and restoring the circuit to operation as an audio amplifier.

Construction

Refer to Figs. 12-3 and 12-4. Switches S1, S2, and S3 are installed on the plastic case below the speaker opening. The speaker is simply glued in place behind the grille using contact cement. (Be careful not to get any on the speaker cone.)

The IC amplifier is assembled on a perf board measuring 3½ inches × 1¾ inches. A 14-pin DIP socket is used for the IC to prevent it from being damaged by excess heat during soldering. Placement and connection of other parts are also shown in the photographs.

The two battery holders fit on either side of the case, above the perf board. When in position, they are held in place by a block of polyurethane foam. (A piece of household sponge will do nicely.)

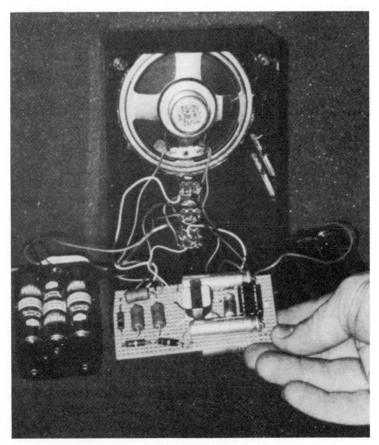

Fig. 12-3. Amplifier is assembled on a perf board.

Fig. 12-4. Amplifier circuit and batteries fit neatly inside of case.

Securing the back to the case holds everything firmly in place.

Ordinary lamp cord or speaker wire can be used to wire the master station to the slave. Runs up to a few hundred feet are tolerable, although output will be reduced by the wire resistance if the run is excessive. Try a higher-impedance slave speaker if a long run is needed.

The slave speaker can be any type: a paging trumpet, a "satellite" speaker (Fig. 12-5), or a simple speaker in a case like that used at the master station. The amplifier will work well with speakers of 8- or 16-ohm impedance.

Fig. 12-5. Satellite speaker can be used as slave speaker.

A parts list for the Squawk-N-Talk Intercom is given in Table 12-1.

Table 12-1. Squawk-N-Talk Intercom Parts List

R1, R2, R3—5600-ohm, ½-W, ±10% resistor
C1, C2, C3—200-μF, 25-V electrolytic capacitor
C4—5-μF, 25-V electrolytic capacitor
C5, C6, C7—0.1-μF, 80-V miniature Mylar capacitor
IC1—audio amplifier ULN2280B (Sprague) or LM280N (National)
S1—miniature spst push-button switch
S2—miniature spdt toggle switch
S3—miniature dpdt toggle switch
T1—8-ohm to 10,000-ohm audio transformer
IC socket—14-pin, dual in-line
Speaker—8-ohm, 3″ diameter, 0.3-W
Misc—2 penlight battery holders (each holding 6 AA-size alkaline cells), plastic case 3½″ × 6″ × 2″, 1¾″ × 3″ perf board

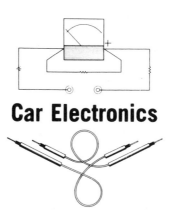

Car Electronics

Electronics is the car-owner's best friend! It brings precision and reliability to the tasks of ignition and battery-voltage regulation in all late-model cars. Electronics has added new dimensions to car-borne entertainment, far enhanced driving pleasure, as well as provided vital communications for the stranded traveler whose car has broken down.

Yet, with all these improvements, there are still many more ways that electronics can enlarge the convenience of driving, give you a maintenance "edge" to keep costs in check, and add to the security of ownership of your prized vehicle.

Gathered here are projects you can enjoy building and using for years to come. Each serves a worthwhile purpose that will help you get the most enjoyment out of your car or recreational vehicle.

While you build these projects, you will also find your experience growing in the details of applying electronics to the harsh environment of motor vehicles. Remember that a good part of every car's lifetime is spent in chilling cold or baking heat, and that severe transients are common even on electronically regulated automotive 12-volt power sources. Each project described has been tested and optimized for the car environment. Given care and attention to detail in construction, each project should reward you with a large measure of satisfaction and enjoyment in use.

Here is an instrument that will help you get top-notch efficiency from your car engine. You can use it to measure rpm for accurate tune-up, timing, idle, and fuel-mixture adjustments. You can diagnose faulty automatic choke operation, slippage in automatic transmissions and clutch assemblies, and excessive fuel consumption due to improper carburetor settings. But, more than this, you can use the Audio Tach on the highway to "program" gear shifting for maximum performance or to warn when engine "over-rev" is approaching.

The Audio Tach (Fig. 13-1) measures engine rpm by counting ignition pulses at the breaker points and displaying its count on a meter that is calibrated in rpm. Apart from this visual display, the Audio Tach has a built-in audible alarm, which can be set to "sound off" when a specific rpm is reached! This audible-alarm feature lets you drive with your eyes on the road, rather than "glued" to the tach.

Fig. 13-1. The Audio Tach.

The Audio Tach features a full-scale measurement capability of 8000 rpm, making it adaptable for use with any 2-, 3-, 4-, 6-, or 8-cylinder engine. It works equally well with two-cycle or four-cycle engines, having either conventional or magneto-type ignition systems. Its layout is not critical, but care must be exercised to use *only* the components specified in the parts list.

The Audio Tach is housed in an inexpensive Bakelite instrument case, measuring 6¼ inches × 3 11/16 inches × 2 inches. All major components are contained on a sheet of perforated phenolic board, measuring 4¾ inches × 3 inches. "Sandwich" construction methods are used to make a compact assembly. The meter, switches, and 2-inch-diameter loudspeaker are installed on a sheet of perforated metal (or perforated phenolic board) cut to exactly 6 inches × 3 inches in order to fit the panel opening of the Bakelite case. The wired component board mounts to the back of the meter by means of the two meter terminal screws. No other supports or hardware is needed. Leads from the component board run to the panel-mounted parts and to a three-terminal barrier-type strip on the back of the case. These terminals are connected to your engine's breaker points and to the positive and negative terminals of a 12-volt dc source, as described later.

How It Works

The schematic of the Audio Tach is shown in Fig. 13-2. The unit consists of two major sections: the *pulse-rate tachometer* and the *audible alarm.*

The pulse-rate tachometer consists of a monostable multivibrator (Q1, Q2) driving a milliammeter, M1. The monostable circuit is like a spring-loaded switch. When triggered by a pulse, it flips to the "on" state and remains there for a specific time, then resets itself automatically. In this application, the monostable circuit generates a series of pulses that are identical in level and duration, even though the pulses produced by the engine breaker points vary in duration with changing speed. Ringing and overshoots in the input pulses are eliminated by a filtering and clipping network (R1, C1, C2, R2, R3, D1, and C3) at the base of Q1. Transistor Q1 is normally off (nonconducting), and transistor Q2 is normally on (conducting). Each input pulse from the breaker points drives Q1 into conduction, supplying a pulse through capacitor C4 to the base of Q2, which drives Q2 off. Transistor Q2 remains off until C4 has discharged through time-trimmer potentiometer R6. During this interval, Q1 is held on by a positive voltage supplied to its base through resistor R4. When C4 is fully discharged, Q2 reverts to its normal on state. The loss of positive voltage at the collector of Q2 causes Q1 to be switched off until the next breaker-point input pulse occurs. Meter M1 is connected as a dc voltmeter in series with diode D2 and scale-adjust potentiometer R8. Each time Q2 is switched off, a voltage appears across M1, causing the pointer to deflect. The rapid on-off switching of Q2 results in a pulse voltage at the collector of Q2 that is directly proportional to rpm. The meter acts to integrate

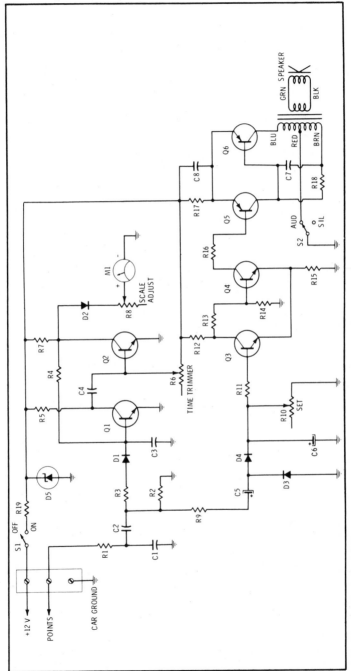

Fig. 13-2. Schematic diagram for the Audio Tach.

these pulses into a fairly steady reading. Diode D2 eliminates the slight saturation voltage present at the collector of Q2 when it is on, improving meter accuracy.

The audible alarm section of the Audio Tach consists of an integrator (D3, D4, C6), a Schmitt-trigger voltage-level detector (Q3, Q4), a transistor switch (Q5), and an audio oscillator (Q6). The integrator is connected to the input filter network and receives the input pulses from the ignition breaker points through capacitor C5. Diode D4 passes the positive portions of each pulse, while D3 clips off negative excursions. Each positive pulse of current passed by D4 charges capacitor C6 in "steps," increasing the voltage across C6 in proportion to engine rpm. Panel-mounted SET potentiometer R10 is in parallel with C6, forming an RC time-constant network. The setting of R10 determines how quickly C6 will reach a given charge. Therefore, R10 can be precisely set so that a specific pulse frequency is required to enable C6 to reach the required charge level. The charge developed across C6 is applied to the base-emitter junction of Q3 through resistor R11. Transistors Q3 and Q4 comprise a modified Schmitt trigger acting as a voltage-level detector. This circuit is normally inactive until the voltage applied to its input reaches a specified level. At that point, the circuit switches on and remains on until the voltage decreases below the trigger level. Normally, Q3 is off and Q4 is on. Collector current for Q4 is supplied through the emitter-base junction of pnp transistor switch Q5 and resistor R16. As long as Q4 is on, Q5 is also held on, shunting the emitter-base junction of audio oscillator Q6. Hence, the audio oscillator stage is squelched under normal conditions. However, when engine rpm reaches a desired level, the voltage across C6 and R10 triggers transistor Q3 on, regeneratively turning off transistor Q4. When Q4 turns off, Q5 is also turned off, and its emitter-collector resistance jumps to a high value. When this happens, Q6 is biased on and commences to oscillate due to the feedback provided to its base through transformer T1. RC network C7-R18 in the base circuit of Q6 sets the oscillation frequency in the range of 500 to 600 Hz. Emitter-current limiting is provided by resistor R17, which is bypassed for audio frequencies by capacitor C8. Transistor Q6 delivers an audio signal at a level of approximately 50 milliwatts to speaker SK1 through the 8-ohm secondary winding of transformer T1.

The tone output from the speaker alerts the driver that the desired rpm level has been reached. Shifting to a higher gear or decelerating will result in lower engine rpm. When the breaker-point input pulses no longer reach the threshold level of the Schmitt trigger, the circuit reverts to its normal state. Transistor Q5 is turned on again and the audio oscillator, Q6, is biased off.

Power for operation of the Audio Tach is obtained from the 12-volt car battery or from an external 12-volt battery. To ensure accuracy, voltages supplied to the monostable multivibrator and Schmitt trigger are regulated at 9.1-volts dc by zener diode D5, in conjunction with resistor R19.

Construction

Assembly of the Audio Tach is fairly straightforward. If you wish to duplicate our version, follow the layout shown in Fig. 13-3. Cut the front panel from perforated sheet metal or phenolic board material. The 2 9/16-inch-diameter hole for meter M1 can be cut by using a "nibbling" tool. Holes for the two switches and SET potentiometer can be drilled and reamed to size. The 2-inch-diameter loudspeaker requires two holes spaced diagonally 2½ inches apart.

When you have cut and drilled the panel, install the set control, switches, and speaker. Next, cut the phenolic component board to size and place it on the back of meter M1 so that it rests on the meter terminals. Mark the terminal hole locations on the board and then locate the holes for transformer T1 to prevent interference in assembly later.

You can now proceed to assemble components on the component board as shown in Fig. 13-3. Wiring is point to point, and no

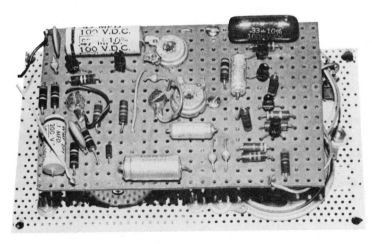

Fig. 13-3. Audio Tach component board.

special precautions need to be observed other than to take care to install the electrolytic capacitors, diodes, and transistors with correct polarity.

When you have finished the wiring, paint the front panel and apply decals for a finished appearance. Next, install the new rpm calibration scale on the face of meter M1. Open the meter by removing the four tiny Phillips-head screws on the back of the meter case. Remove the cover and gently unthread the two screws securing the meter face to the movement. Delicately remove the face, being careful not to damage the meter pointer. A full-size rpm calibration scale (Fig. 13-4) is provided. It can be photocopied and secured over the existing scale by using a drop or two of rubber cement. When it is dry, reinstall the meter face and reassemble the meter. You can now install the meter in the front panel of the Audio Tach.

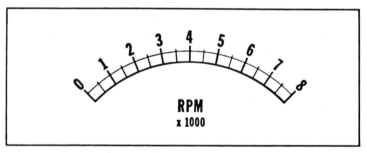

Fig. 13-4. Meter scale for Audio Tach.

Speaker SK1 is held in place by two solder lugs secured with screws, nuts, and washers. The lugs bear against the metal rim of the speaker, holding it securely against the panel. Referring to the schematic in Fig. 13-2, wire the panel-mounted components to the assembled component board. Next, assemble terminal strip TS1 to the rear of the Bakelite case. Drill two holes for mounting the terminal strip and drill three holes opposite the three terminals for connection of leads. Drill the holes carefully to avoid chipping the case. Install the terminal strip and connect one end of a wire lead under each terminal screw. Pass these leads through the holes in the case and connect them to switch S1, resistor R1, and the ground bus. When these steps are completed, make a thorough final check of the wiring and assembly to make certain that there are no wiring errors. When you have done this, you are ready to calibrate your Audio Tach.

Calibration

Your completed Audio Tach must be calibrated for the type of engine with which it will be used. Two calibration methods are given: the first method gives the most accuracy but requires test gear you may not have; the second method yields an instrument of somewhat less accuracy but requires little in the way of test gear. Choose the method that best suits your circumstances.

The first calibration method requires an audio generator capable of producing sine or square waves in the frequency range of 30 to 1200 Hz. If you do not have such a generator, you can take your Audio Tach to a local television or hi-fi service shop that does and have them calibrate it for you. Table 13-1 lists the ignition pulse

**Table 13-1. Ignition Pulse Rate
Versus RPM for Common Engine Types**

Engine Type	Ignition Breaker-Point Pulses per Second							
	Rpm 1000	2000	3000	4000	5000	6000	7000	8000
4-cylinder, 4-cycle	33	66	100	133	166	200	233	266
6-cylinder, 4-cycle	50	100	150	200	250	300	350	400
8-cylinder, 4-cycle	66	132	200	266	332	400	466	532
2-cylinder, 2-cycle (Magneto ignition)	33	66	100	133	166	200	233	266
3-cylinder, 2-cycle	50	100	150	200	250	300	350	400
4-cylinder, 2-cycle	66	132	200	266	332	400	466	532
6-cylinder, 2-cycle	100	200	300	400	500	600	700	800
8-cylinder, 2-cycle	132	264	400	532	664	800	932	1064

rates of most common two-cycle and four-cycle engines, in 1000-rpm steps from 1000 to 8000 rpm. The audio generator simulates the ignition breaker-point pulses for bench calibration. You will also need a 12-volt dc source to power the Audio Tach. Two 6-volt lantern batteries connected in series make an ideal supply. Connect the generator and battery supply to the Audio Tach as shown in Fig. 13-5. Next, referring to the data given in Table 13-1, set the

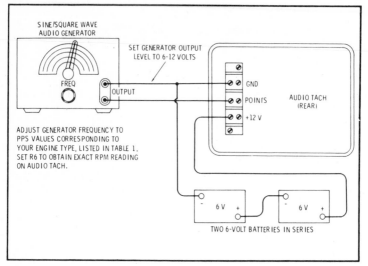

Fig. 13-5. Preferred calibration setup for Audio Tach using an audio generator.

generator frequency to the ignition pulse rate corresponding to 8000 rpm for your engine type. Set the generator output to give at least 8 volts of signal input to the Audio Tach. Now, set switch S1 to the ON position and observe the deflection of meter M1. Adjust time-trimmer potentiometer R6 until M1 just indicates 8000 rpm. Then, set the generator to the frequency corresponding to 1000 rpm and observe the Audio Tach reading. If the meter pointer lies above or below the mark that indicates 1000, set scale-adjust potentiometer R8 so that the meter indicates correctly. Recheck at the frequency corresponding to 8000 rpm and readjust time-trimmer R6 if necessary.

To check the audible-alarm portion of the circuit, set the generator to the frequency corresponding to 5000 rpm for your engine type, and adjust the set potentiometer (R10) until the audio oscillator sounds. Decrease the generator frequency setting until the audio oscillator is silenced; then run it back up to the previous setting. As the Audio Tach meter reaches the 5000-rpm mark, the audio oscillator should trigger again. Upon completion of this procedure, the Audio Tach may be assembled into its case and installed in your car.

The second calibration method enables you to do a fair job of setting up your Audio Tach by using the test setup in Fig. 13-6 instead of an audio generator. In this method, the 60-Hz frequency of the ac power line is used to simulate ignition pulses. While the accuracy of this method is not as precise as the previous method, it is sufficient for an instrument that will be used casually.

Make the setup as shown in Fig. 13-6 and connect the transformer primary to a 117-volt, 60-Hz ac line. Rotate scale-adjust potentiometer R8 for maximum deflection of M1. Do not allow the meter to slam against the pin; just set it for maximum deflection, or to 8000 rpm, if possible. Now, adjust time-trimmer potentiometer R6 for the meter rpm reading corresponding to your type, as shown in Fig. 13-6. If necessary, set R8 in conjunction with R6 to get an exact reading.

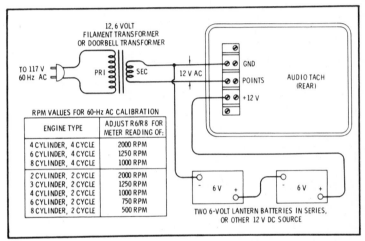

Fig. 13-6. Calibration setup for Audio Tach using the 60-Hz ac-line frequency.

After you have made this adjustment, you can proceed to hook up the Audio Tach to your engine.

Hooking It Up

Connecting the Audio Tach to most engines is simple. For conventional ignition systems, connect a lead between the points terminal on terminal-strip TS1 and the (−) terminal of the engine ignition coil, as shown in Fig. 13-7. Connect the gnd terminal of TS1 to any convenient ground screw on the chassis or the engine block. Attach the lead from the +12-V terminal of TS1 to any convenient terminal on the auto fuseblock (usually located under the dash on the driver's side). The fused instrument or accessory terminals are ideal for this connection. If you wish to make a permanent installation in an auto, truck, or boat, you need only run the points lead into

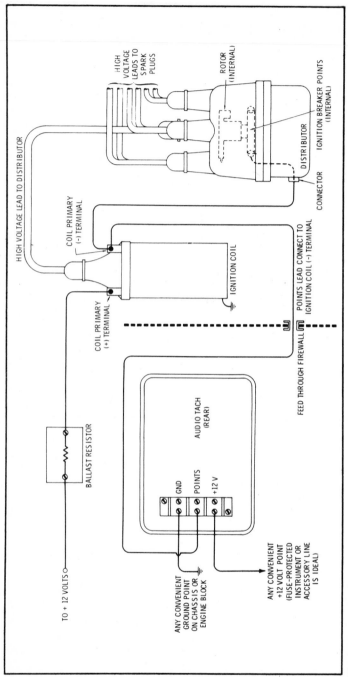

Fig. 13-7. Audio-Tach hookup for a conventional ignition system.

the engine compartment. The ground and +12-V leads can be connected in the driving compartment. (An ideal feedthrough point for the points lead is the grommeted speedometer hole, usually located on the firewall just below the dashboard on the driver's side of most cars.) Where an electronic ignition system is used, connect to the + terminal of the ignition coil, or follow the manufacturer's hookup that is recommended for a tachometer.

For operation with magneto-type ignition systems (used in outboard motors, tractors, and motorcycles), you will have to make some provision for power to operate the Audio Tach. Two 6-volt lantern batteries connected in series make a long-life 12-volt supply that can be stashed away in any convenient location. The batteries can be connected by reasonable-length leads to the +12-V and ground terminals of the Audio Tach.

Magneto ignition systems use a built-in generator instead of the battery found in conventional ignition systems. This built-in generator develops a moderate dc voltage that is interrupted by the ignition breaker points and stepped up to spark-plug firing potential through a secondary coil surrounding the generator coil. In most engines employing magneto ignition, you can attach a lead directly to the ungrounded breaker-point terminal (the terminal connected to the condenser lead) and bring it out to the Audio Tach.

In Chrysler outboard engines, the engineers have thoughtfully provided a tachometer lead (color-coded white) as part of the standard harness assembly. This lead comes directly from the breaker points and should be connected to the points terminal on the Audio Tach.

If you have any doubts about connecting the Audio Tach to your ignition system, consult your engine overhaul manual or the service department of a local dealer agency for details.

Using the Audio Tach

You will find the Audio Tach a real help in tuning up your car's engine. Antipollution devices call for timing and carburetion adjustments to be made at specific engine rpm. You can set and hold rpm exactly as specified by using the Audio Tach meter. Also, you can check operation of your engine's automatic choke. Engine rpm should reduce with warm-up, and the Audio Tach meter will show this far more accurately than you can detect by ear. Faulty ignition components will also show up on the Audio Tach meter. Erratic meter readings may indicate bouncing points, a defective coil or condenser, or excessive wear in the distributor cam shaft. Automatic transmissions can also be checked with the Audio Tach. Since shift points of the transmission are directly related to engine rpm, slippage in the transmission can be detected by watching the meter pointer and seeing if it exceeds the shift rpm specified by the manufacturer. Engine idle speed is another critical adjustment in which the Audio Tach can help a lot. If your engine idles too slow, it may stall easily. An idle speed that is too fast means wasted fuel, excessive strain on the transmission and brakes, and the annoyance

Table 13-2. Audio Tach Parts List.

C1, C3—.01-μF, 600-Vdc disc capacitor
C2—.1-μF, 200-Vdc capacitor
C4—.47-μF, 100-Vdc capacitor
C5—10-μF, 100-Vdc electrolytic capacitor
C6, C8—10-μF, 25-Vdc electrolytic capacitor
C7—.33-μF, 100-Vdc capacitor
D1 to D4—general-purpose silicon diode type 1N5060
D5—9.1-V, 1-W zener diode (Motorola HEP Z0412, Radio Shack 276-562, or equiv)
M1—0-1 milliampere meter (Lafayette 99F50874, Radio Shack 270-1752, or equiv.)
Q1,Q2—npn transistor, type 2N3414 or equiv (Radio Shack 276-2009)
Q3, Q4—npn transistor, type 2N5172 or equiv (Radio Shack 276-2009)
Q5, Q6—pnp transistor, type 2N2907 or equiv.
R1, R2, R3, R9, R11—1000-ohm, ½-W ± 10% resistor
R4—15,000-ohm, ½-W ± 10% resistor
R5—1200-ohm, ½-W ± 10% resistor
R6, R8—10,000-ohm potentiometer (Mallory type MTC-14L4 or equiv)
R7—820-ohm, ½-W ± 10% resistor
R10—25,000-ohm potentiometer (IRC type PQ11-120 or equiv)
R12, R16—4700-ohm, ½-W ± 10% resistor
R13—2200-ohm, ½-W ± 10% resistor
R14—6800-ohm, ½-W ± 10% resistor
R15—150-ohm, ½-W ± 10% resistor
R17—100-ohm, ½-W ± 10% resistor
R18—7500-ohm, ½-W ± 10% resistor
R19—220-ohm, ½-W ± 10% resistor
S1, S2—miniature spdt toggle switch (Alcoswitch type MST-105D or equiv)
SK1—2-inch-diameter, 0.25-W, 8-ohm speaker
T1—transistor output transformer, 1000-ohm ct primary, 8-ohm secondary (Argonne type AR-137, Radio Shack 273-1380, or equiv)
TS1—three-terminal barrier strip (Cinch-Jones type 3-140, Radio Shack 274-657, or equiv)
Misc—Instrument case, 6¼" × 3¾" × 2" (Radio Shack 270-627 or equiv); perforated panel, 4¾" × 3" perforated phenolic component board; knob; hookup wire; solder.

of excessive "creep." You can precisely set idle speed with the Audio Tach and, by teaming it up with a vacuum gauge, you can make a professional fuel-mixture adjustment for all-around best performance and economy.

The audible-alarm feature of the Audio Tach can be used anytime you want an indication of the meter reading but cannot divert your attention to watch it. This is especially helpful at night, when your eyes should not leave the road, even for an instant.

Drivers of cars with manual transmissions gain extra benefits from the Audio Tach. If you know the manufacturer's specified rpm for maximum torque output from your engine, you can set the Audio Tach to sound when that rpm is reached. You can then shift and ride the rpm up to maximum torque again all without once taking your eyes off the road ahead!

For "over-rev" protection, just set the audible alarm to sound about 500 rpm below the engine red-line rpm. This setting will guarantee your being warned in the event that rpm is becoming excessive.

A parts list for the Audio Tach is given in Table 13-2.

14. BATTERY-CONDITION CHECKER

Each of us face a "moment of truth" when we turn the key in the ignition switch for the morning start-up. The car battery is given a rigorous exam by the voracious starting motor. Battery current of nearly 400 amperes is needed to crank a cold, high-compression engine off dead center. If the battery passes, you start; if not, you sit and listen to the disheartening "clickety-click-click" of the solenoid as it vainly makes and breaks the circuit to the deceased battery.

Not all of us are as sensitive as we should be to the somewhat subtle indications of a failing battery. It takes a keenly developed sense of hearing to detect a slower-than-normal cranking speed, a slight hesitation in the cranking action, or a somewhat longer-than-usual cranking period. The instruments used to indicate the condition of the electrical system in most vehicles are barely any help, either. The widely misunderstood charging ammeter gives ambiguous indications. It reads zero-center for a fully charged battery—but you get the *same* zero indication if the alternator belt is slipping or the regulator has failed. The "idiot light" detects only complete failure of the charging system; it cannot determine the battery's capacity to deliver energy.

The 12-volt lead-acid storage battery that powers the starting system of an automobile is often subjected to harsh use. Low temperature extremes sap the battery's capacity, and frequent start-ups, little actual driving, or defects in the charging system can keep it constantly undercharged. It is little wonder that *battery failure* ranks so high on the list of problems that car owners suffer.

Yet, battery failure does not occur unannounced. The conditions that will lead to a "no-start" if left unchecked are usually detectable days or even weeks in advance. All that is needed is the *means* of detecting them—*while there is still time to do something about them!*

Engineers have long recognized that an accurate test of the condition of a lead-acid battery can be made by measuring the voltage across its terminals *while the battery is actually supplying current to a load.* It is generally agreed that the terminal voltage of a 12-volt automotive battery should not fall below 9.2 volts when initially loaded by the starting motor. The *automatic battery-condition checker* shown in Fig. 14-1 allows the driver, while comfortably seated, to check his battery in seconds as part of the normal morning start-up routine.

The complete circuit is contained in a cigarette-lighter accessory plug, usually used to tap power for accessories such as lights from the vehicle's cigarette-lighter socket. The plug body has two

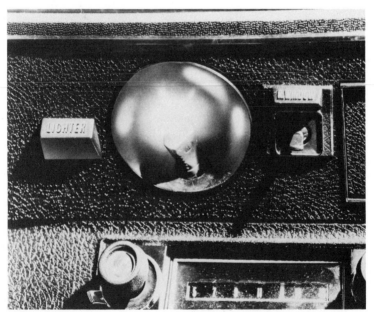

Fig. 14-1. Battery-Condition Checker installed in cigarette-lighter socket.

terminals—a blunt positive terminal at the tip and a flat negative terminal along the side. An indicator lamp is mounted at the rear end of the plug body. The checker is designed to slide into the cigarette-lighter socket and thereby establish contact with the battery through the vehicle wiring.

A simple voltage-sensitive switching circuit is incorporated entirely into the body of the checker. This circuit is designed to discriminate between the terminal voltage under load that represents a battery in good condition and the voltage that represents a discharged or deteriorated battery.

How It Works

The checker is inserted into the cigarette-lighter socket. Immediately, the lamp glows and remains on. The driver now cranks and starts the engine while observing the checker lamp. Should the lamp extinguish during cranking, the voltage across the battery has dropped *below* the acceptable norm for an average battery under the load imposed by the starting motor. Should the lamp go out, even if the car starts, trouble may be at hand. Of course, if the battery is good, the lamp remains glowing during the entire cranking interval. After starting, you simply unplug the checker and store it in the glove compartment. Although a daily check is informative, a weekly check should be adequate to detect the onset of battery deterioration.

The input circuit (as shown in Fig. 14-2) comprises a silicon diode and a zener diode. Diode D1 has an approximate (but virtually constant) turn-on voltage of 0.6 volt. The zener diode, D2, has an avalanche voltage of 7.5 volts (also constant). Thus, current has no path through the series diodes until the potential difference is greater than 8.1 volts. The output voltage at the junction of D2 and R2 is applied to transistor Q1 and Q2, which are connected as a Darlington transistor switch. This compound transistor does not

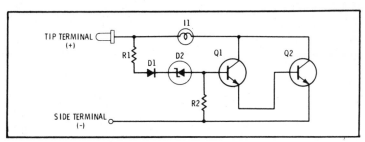

Fig. 14-2. Battery-Condition Checker schematic.

conduct until the potential difference between its base and emitter exceeds 1.2 volts. Only then does this transistor combination act as a closed switch so that current may flow through indicator lamp I1. This means that the voltage which *must* be present at the input of Q1 cannot be *less* than 9.3 volts (0.6 V + 7.5 V + 1.2 V), if the lamp is to be switched on. With an input voltage above 9.3 volts, the circuit functions as it would at 9.3 volts (lamp on). During cranking, if the battery voltage falls *below* 9.3 volts, forward bias is removed from the base of Q1 and the Darlington switch is turned off. Thus, the lamp goes out, and you have an indication that the battery has failed under load!

There are no controls, and indicator interpretation is simple. If the battery is good, the lamp is on; if the battery is bad, the lamp is off. The automatic battery-condition checker is readily transferred from car to car, as well as to a boat or snowmobile.

Construction

The seven components plus indicator light that make up the circuit all fit snugly *inside* a standard cigarette-lighter accessory plug (Figs. 14-3 and 14-4). First, drill out the rivet that holds the two halves of the plug together and remove the small separating struts at the rear of the plug. Next, hold the two halves of the plug firmly together and drill (or file) a 5/16-inch-diameter hole to a depth of ⅜ inch for the pilot light.

The pilot-light assembly is 15/16 inch long and must be shortened to fit in the hole. Cut away 3/16 inch for an overall length of ¾ inch. To make room for the pilot-light assembly, the two metal strips are bent around the two plastic posts which support them.

Fig. 14-3. Assembled Battery-Condition Checker.

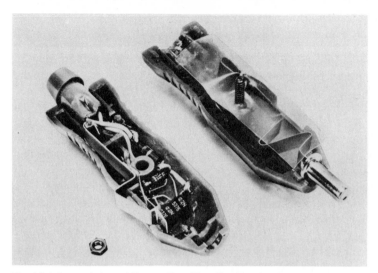

Fig. 14-4. Internal view of Battery-Condition Checker showing component layout.

Wire the transistors, diodes, and resistors inside the plug as shown in Fig. 14-4. The whole assembly nestles nicely inside one-half of the plug after some of the struts are removed with a pair of long-nose pliers. Be careful that none of the connections short. To prevent this, we used a few small pieces of clear plastic separators cut from the blister package of the accessory plug. The pilot-light assembly can be held in place with a dab of clear cement. A ¾-inch-long screw and nut replace the rivet and hold the two halves firmly together. Neither the screw nor the nut may extend beyond the body of the plug. If they do, the plug will not fit into the cigarette-lighter socket.

A parts list for the Battery-Condition Checker is given in Table 14-1.

Table 14-1. Battery-Condition Checker Parts List

D1—general-purpose silicon diode (Motorola HEP R0052 or equiv)
D2—7.5V, 1-W ± 5% zener diode (Mallory ZB7.5B or equiv)
Q1, Q2—npn silicon transistor, type 2N5376 or equiv
R1—120-ohm, ½-W ± 20% resistor*
R2—4700-ohm, ½-W ± 20% resistor*
I1—12-V, 150-mA max incandescent pilot light (Industrial Devices 2990D1-12V or equiv)
Cigarette-lighter accessory plug (Radio Shack 274-331 or equiv)

*If available, use ¼-watt resistors for easier fit.

Reliable as automotive lamps are, they can *still* burn out at unexpected moments. Perhaps the most dangerous failure is the loss of the rear brake light, since this type of failure robs your car of two important signaling functions: it reduces the warning to the driver behind you that your car is braking, and it sharply curtails your ability to signal a turn. Of course, the degree of severity of the failure is related to the number of lamps in your car's rear-end signaling system. If your car has only *one* stoplight for each side, a lamp failure can be an open invitation to a deadly collision!

The most insidious factor in brake-lamp failures is that the filament opens up quietly, and in a position that cannot be seen from the driver's seat. Occasionally checking the lamps at a service station with the aid of an attendent may help, but it is no guarantee that a brake lamp won't fail two minutes after you have driven away. The brake-light monitor shown in Fig. 15-1 gives you a dependable indication of whether or not your car's most vital signal lamps are functioning properly. Should any lamp become inoperative, a light will instantly flash on in the monitor to warn you of the hazardous condition.

Fig. 15-1. Brake-Light Monitor.

How It Works

The most important part of the circuit is a tiny reed relay to which a second winding, consisting of 6 turns of No. 14 enamelled wire, has been added. As shown in Fig. 15-2, this added winding is connected in series with the lead that goes from the stoplight switch to the brake lamps, so that the full lamp current flows through this winding enroute to the lamp filaments. The lamp current produces a magnetic field that is proportional to the current required by the filaments. In the case of a two-lamp stoplight system, this current would be from 4 to 5 amperes. This rather respectable amount of

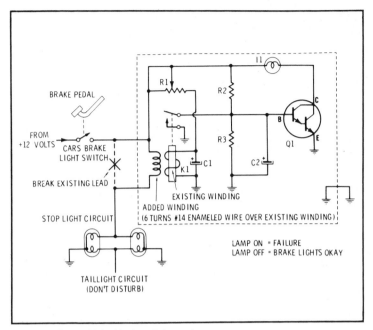

Fig. 15-2. Brake-Light Monitor Schematic.

current produces a fairly strong magnetic field, which is *added* to the magnetic field generated by the existing winding on relay K1. The field strength of the existing winding is adjusted by R1 so that it is just sufficient to close the reed contacts. Hence, the magnetic field produced by the stoplight current flowing through the additional winding is summed with the field of the existing winding every time you step on the brakes. If any one of the lamps is inoperative (either because it is burned out or because of poor electrical contact) there will be proportionately less current through the added 6-turn winding and the reed relay will not close.

Capacitor C2 and resistor R2, together with Darlington transistor Q1, form a one-second time delay circuit. The collector Q1 is connected through indicator lamp I1 to +12 volts. At the instant that the brake pedal is depressed and +12 volts is applied to resistor R2, the delay is started. After one second, capacitor C2 is charged and transistor Q1 will turn indicator lamp I1 on *unless* the reed relay is closed. If the relay closes, the base of transistor Q1 is connected to ground and the transistor cannot turn on. As you can see, current flow corresponding to "good brake lamps" automatically disarms the circuit controlling the indicator lamp.

The directional-light circuit causes one *rear lamp* to flash, but the monitor delay circuit keeps the failure lamp "off" during the short period of time that the stoplight circuit is opened for blinking action. If the delay circuit were not incorporated, the brake-light monitor lamp would flash every time you signaled a turn. It is possible that the directional lights will flash only for a very brief period of time; this may be the case when you have a defective flasher or use incorrect bulbs. In either case, the "off" period would be beyond normal limits. Thus, if the circuit is interrupted for more than one second, the brake-light monitor will flash regularly (or occasionally) whenever your turn signals are on, thereby warning you of a defect in your directional-light circuit.

Construction

The unit is housed in a 4-inch × 2⅛-inch ×1⅝-inch aluminum minibox. As shown in Fig. 15-3, the failure indicator lamp is mounted on one face of the U-shaped channel, and the barrier terminal strip is mounted on the rear face. The left terminal on the barrier strip goes to +12 volts through the brakelight switch. The center terminal goes to the stoplights, and the third terminal goes to chassis ground. All the electronic components are mounted on one side of a 2-inch × 3½-inch piece of perf board. For ease of assembly, flea clips are used to make the connections on the perf board. The perf board is held in place by two ¾-inch screws that go through one-half-inch spacers that keep the board a proper distance from the chassis to prevent shorts.

Fig. 15-3. Internal view of Brake-Light Monitor showing component layout.

Calibration

The circuit sensitivity needs to be adjusted only one time. Step on the brake pedal so that the stoplights go on. Simultaneously adjust potentiometer R1 from full resistance (lamp on) until the reed relay just closes and indicator lamp I1 goes off. To simulate lamp failure, reach into the trunk and unplug one stoplight assembly from the frame of the car, making sure it does not touch the metal body of the car. Now, when you step on the brake pedal again, the indicator light should go on after one second, signaling that the lamp circuit is not operating properly. If the indicator lamp does not go on, increase R1 slightly until the lamp just goes on. Then, reinsert the stoplight assembly previously removed. Pump the brake pedal once again. This time the warning lamp should not go on, indicating that the stoplight circuit is all right. For proper operation of the monitor, it is necessary that the fields of both relay windings *aid* each other. If it is not possible to calibrate the monitor by adjusting the potentiometer, simply reverse the two lead connections to the added-turn winding so that the direction of its current flow is reversed. This will solve your problems and proper adjustment will then be possible.

Table 15-1. Brake-Light Monitor Parts List

R1—25,000-ohm potentiometer (Mallory MTC253L4 or equiv)
R2, R3—100,000-ohm, ½-W ± 10% resistor
C1—20-μF, 50-Vdc electrolytic capacitor
C2—10-μF, 25-Vdc electrolytic capacitor
Q1—npn silicon Darlington transistor, type 2N5306 or equiv
K1—reed relay, 12-Vdc, 120-ohm coil (Magnecraft W101 MX-2)
I1—12-V pilot light (Industrial Devices 2990D1-12V or equiv)
Misc—4″ × 2⅛″ × 1⅝″ aluminum minibox (Bud CU-2102A, Radio Shack 270-239, or equiv); three-terminal barrier strip (Cinch-Jones type 3-140 or equiv); 2″ × 3½″ perf board; ½″ spacers (2); screws; flea clips

16. COMPUTALARM

While you are reading this chapter, an automobile is being stolen somewhere in the United States. Every minute of every day, someone's car is driven away without the owner's knowledge. This amounts to more than 500,000 cars being stolen every year. The chances that your car will be stolen are increasing every year. Yet, only 30 percent of stolen cars are taken by professional car thieves. Most car theft is the work of moonlighting amateurs. Even if a car is recovered (a good number are), chances are it will be stripped-down or wrecked. Car insurance offers some relief but generally does not cover all costs, inconvenience, and loss of property or special equipment installed in your car. Build the all-electronic Computalarm shown in Fig. 16-1 for a very nominal one-time expense, and you can protect the hood, trunk, and passenger compartment of your car against assault by any thief.

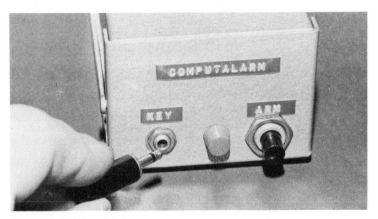

Fig. 16-1. The Computalarm.

Many of the automobile burglar alarm systems on the market today are either relatively expensive, difficult to install, or inadequate in their protection. Some put a prolonged and undue strain on the battery as well as on the ears and nerves of people within a block or more. Such alarms usually continue their shrieking action for a far longer period than is necessary to scare away the would be thief or attract the attention of a policeman. Also, there is not much point to protecting the car with one of these types of alarms if it means that the battery will be exhausted when you return!

This is not the case with the Computalarm. The alarm remains in action for a sufficiently long period of time to scare away the would-be thief. It then turns itself off and automatically rearms. It is ready to do its job again, without having driven all bystanders out of their minds and without having drained the battery.

The Computalarm protects as many points of entry to the car as you wish. It has several distinct operating features:

1. The circuit has a built-in, self-arming feature. The driver turns off the ignition, presses the arm button on the Computalarm, and leaves the car. Within 20 seconds the alarm arms itself—all automatically!

74

2. The circuit will then detect the opening of any monitored door, the trunk lid, or the hood on the car. When triggered, the hood lid is closed, the circuit remains activated.
3. Once activated, the circuit remains dormant for 10 seconds. When the 10-second time delay has run out, the circuit will close the car's horn relay and sound the horn in periodic blasts (approximately 1 to 2 seconds apart). For a period of one minute. No burglar is likely to wait around that long with the alarm on. Then the Computalarm automatically shuts itself off (to save your battery) and rearms. Of course, if a door, the trunk lid, or the hood lid remains ajar, the alarm circuit retriggers and another period of horn blasts occurs.
4. The Computalarm has a "key" switch by which the driver can disarm the alarm circuit within a 10-second period after he enters the door. The key switch consists of a closed circuit jack, J1, and a mating miniature plug that is small enough to carry on your key chain (Fig. 16-2). The plug fits into a similar jack on your key chain and is merely removed from that jack and plugged into J1 on the Computalarm to disarm the alarm, so that you alone can enter the car without the horn sounding.

Fig. 16-2. Key for the Computalarm.

How the Circuit Operates

The courtesy-light door switches are used to protect the doors, and single-pole single-throw microswitches are installed under the hood and trunk lid of the car. These switches are installed so that when a door, hood, or trunk lid is closed, the switches remain open. When any portal is opened a switch closes and triggers the circuit. (You can also experiment with a variety of other switch types like mercury switches or acceleration-sensitive switches. Thus, if someone jostles the car or attempts to move it or jack it up to steal your wheels, the alarm circuit will be activated.)

When a switch closes, it applies a ground to one end of resistor R1, R2, or R3 at the input to transistor Q1, which is normally not conducting. (See Fig. 16-3). When one of the resistors is grounded, Q1 turns on and the collector voltage at R9 jumps up the +12-volt level of the battery. This voltage causes zener diode D2 to conduct and develop about 2 volts across resistors R4 and R5. This voltage

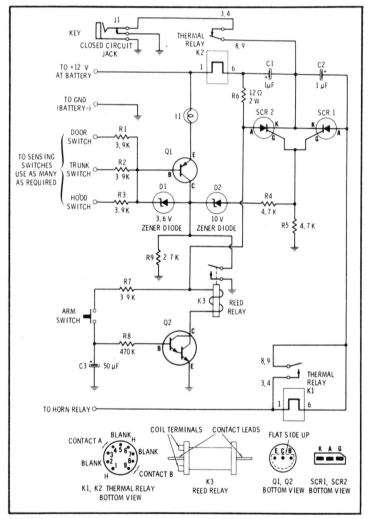

Fig. 16-3. Schematic diagram for the Computalarm.

is sufficient to turn on SCR1 and SCR2, both of which are sensitive silicon-controlled rectifiers. When these SCRs turn on, they latch and allow current to flow through the heater windings of thermal relays K1 and K2. (Once an SCR has been turned on, it will continue to conduct as long as a source of current is applied to its anode and the return-path circuit remains complete.)

This is how the Computalarm latching action works. When the heaters of relays K1 and K2 are energized, they commence to warm bimetallic contacts. Relay K1 is a normally open thermal relay, while relay K2 is a normally closed thermal relay. When energized, relay K1 will close itself within 10 seconds. Similarly, relay K2 will open within 60 seconds. The cathodes of SCR1 and SCR2 go to ground through the closed contacts of thermal relay K2 and through jack J1, which is closed whenever the plug (key) is not in it. Within 10 seconds after SCR1 is gated on, the contacts of thermal-relay K1 close. This grounds one end of the car's horn relay, causing the horns to sound. To make the horn operate in blasts, relay K1 is arranged so that when its contacts close, its heater is momentarily shorted, causing it to cool and open the bimetallic contacts. The low heater current through the horn-relay winding is not sufficient to cause the horn to blow, but as the heater warms once more, the contacts of K1 close again. The cycle repeats in a period of about every 2 seconds. While relay K1 is operating the horn circuit, the bimetallic contacts of relay K2 are receiving a constant warming from its heater.

After somewhat more than 1 minute, the contacts of relay K2 are sufficiently warm to open. This interruption causes both SCRs to commutate, which stops the horn from sounding and automatically resets the circuit.

If a door, trunk lid, or hood remains ajar, a ground will still be present on one of the resistors (R1, R2, or R3) at the base of transistor Q1, and the cycle will repeat. If the door, trunk lid, or hood has been slammed shut, the circuit will return to the standby condition when the SCRs commutate and the horn will stop sounding.

Since the Computalarm responds to a switch closure every time a door is opened, some means must be provided to allow the driver or passenger to leave the car without falsely actuating the alarm. This is the function of the delay circuit consisting of transistor Q2 and reed relay K3. Transistor Q2 is a very sensitive amplifier which can be made to operate like a switch that remains closed for a period of time and then opens by itself. The input circuit to Q2 consists of capacitor C3 and resistor R8.

When the driver and passengers want to leave the car, the driver merely waits until all passengers have left the car and closed their doors. Next, he removes the plug (key) from jack J1 and depresses the arm push-button switch. When he does this, the capacitor C3 in the base circuit of transistor Q2 is charged quickly to +12 volts. This turns on Q2, which pulls in the contacts of reed relay K3, grounding the collector of transistor Q1.

Indicator light I1 is now turned on to indicate that the circuit is getting ready to arm itself. After about 20 seconds, which gives the driver sufficient time to leave the car, the light goes out and the circuit is armed. The next time a door is opened, the entrant has 10 seconds in which to insert the correct size "key" into jack J1 or the alarm will sound the horn.

Construction of the Computalarm is simple. As shown in Figs. 16-4 and 16-5, all components with the exception of relays K1 and K2 are mounted on a 2½-inch × 5-inch piece of perf board. The sockets for K1 and K2 are held by screws to one side of the U-shaped portion of the aluminum minibox. The arming push button, indicator light L1, and jack J1 are mounted on the opposite side. The wires to the protective switches, the battery, the horn relay, and ground are passed through a rubber grommet.

The Computalarm can be mounted at any convenient location under the dash, or even in the glove compartment. Choose a spot in your car that is best suited for you.

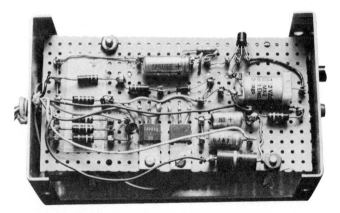

Fig. 16-4. Internal view of Computalarm showing component layout.

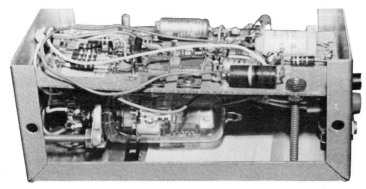

Fig. 16-5. Internal view of Computalarm showing mounting for thermal relay K1.

In some cars, the horn may be disconnected relatively quickly because of the use of quick-disconnect cable plugs that are used in most modern automobiles. To thwart any quick disconnecting, we would suggest that these leads be securely taped to the horn terminals. Better protection would be afforded by installing a sheet metal barrier to block access to horns and by using armored cable to protect the horn wiring against attack.

A parts list for the Computalarm is given in Table 16-1.

Table 16-1. Computalarm Parts List

C1, C2—1-μF, 25-Vdc electrolytic capacitor
C3—50-μF, 25-Vdc electrolytic capacitor
D1—3.6-V, ¼-W zener diode, type 1N757A or equiv
D2—10-V, ¼-W zener diode, type 1N765 or equiv
I1—12-V, 100-mA pilot lamp (Industrial Devices 2990A1-12V or equiv)
J1—miniature closed-circuit phone jack (Switchcraft 42A or equiv)
K1—normally open, 10-second delay thermal relay (Amperite 6N010T or equiv)
K2—normally closed, 60-second delay thermal relay (Amperite 6C60T or equiv)
K3—spst, normally open, 12-V reed relay (Magnecraft W101MX-2 or equiv)
P1—miniature phone plug for jack J1
Q1—pnp silicon transistor (Motorola HEP S0012 or equiv)
Q2—npn silicon Darlington transistor, type 2N5306 or equiv)
R1, R2, R3, R7—3900-ohm, ½-W ± 10% resistor
R4, R5—4700-ohm, ½-W ± 10% resistor
R6—120-ohm, 2-W ± 10% resistor
R8—470,000-ohm, ½-W ± 10% resistor
R9—2700-ohm, ½-W ± 10% resistor
S1—spst, normally open push-button switch
SCR1, SCR2—silicon-controlled rectifier (General Electric type C106Y1, or Radio Shack 276-1067, or equiv)
Misc—9-pin miniature tube socket for K1 and K2; 2½″ × 5″ perf board; 5¼″ × 3″ × 2⅛″ aluminum minibox (Bud CU-2106A, Radio Shack 270-238, or equiv); sensing switches as required.

Suppose that a light drizzle or snow is falling, or that the car ahead of you is spraying fine mist and road scum onto your windshield after a snow or rainfall. You turn on your windshield wipers, but, more often than not, the wipers do not move slowly enough to match the aggravatingly slow rate of accumulation on the windshield. Soon, the wipers begin to bounce and squeak. In frustration, you reach for the windshield wiper switch and turn it on and off, to suit your needs. This maneuver seems to do the job, but it also takes your attention away from the road and manages to keep one hand continuously occupied. In general, it makes for unsafe and unpleasant driving.

With the electronic Windshield-Wiper Controller shown in Fig. 17-1, you can have complete speed control over your car's windshield wipers. They can be slowed down to any rate that suits your needs—even down to four sweeps per minute.

Fig. 17-1. Windshield-Wiper Controller.

How It Works

The Windshield-Wiper Controller has two principal circuits: the rate-determining circuit and the actuator (Fig. 17-2). The rate-determining circuit consists of a unijunction transistor connected as a free-running oscillator. The rate of the pulse output at base 1 (B1) of the unijunction transistor (Q1) is determined by the combination of C1 and the sum of R1 and R2. When the voltage across C1 reaches a predetermined value, Q1 suddenly conducts the C1 discharges

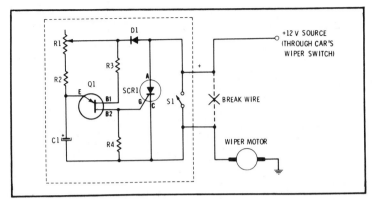

Fig. 17-2. Schematic diagram of the Windshield-Wiper Controller circuit.

through R4, producing a positive pulse. This pulse is applied to the gate (G) of the silicon-controlled rectifier which is the actuator. Placed in series with the +12-volt feed to the wiper motor, the SCR "latches" when triggered, turning on the wiper motor. At the end of each operating cycle, the return (parking) switch in the wiper mechanism removes the dc voltage or reverses the motor connections, which turns off the SCR and brings the wiper blades to rest. However, as long as S1 is open, capacitor C1 will again begin to charge until the unijunction transistor conducts once more and the cycle repeats. Connection of the controller in no way affects the usual operation of the car's wiper switch. When the controller is turned off, switch S1 closes and shunts the SCR, bypassing the controller circuit for normal operation of the windshield wiper.

Operation of the controller can easily be checked with a 12-volt bulb. Place the bulb in series with one of the leads and connect the controller to a 12-volt source, observing the correct polarity. The bulb will turn on after a delay determined by the setting of R1. To turn it off and check for the next cycle, the voltage source must be momentarily removed.

CAUTION: Do not connect the controller directly across a 12-volt source, since this will immediately burn out the SCR. A load must be connected in series with the controller and the power source.

Construction

The controller is housed in a 3¼-inch × 2⅛-inch × 1⅝-inch aluminum minibox, and all of the circuit components are mounted on a 1¾-inch × 2¾-inch perf board, which is held in place by an extra nut on the control. A cardboard "antishort" shim is used to prevent any possibility of the components' shorting to the aluminum case. The SCR is mounted on a small L-shaped metal bracket. (The SCR is a heavy-duty type, and its operating cycle is short enough to make the use of a heatsink unnecessary.) The parts layout is shown in Fig. 17-3. The wire leads to the wiper switch should be at least No. 16 stranded wire and no longer than necessary, since a current of several amperes is being switched.

Fig. 17-3. Internal view of the Windshield-Wiper Controller.

The controller can be mounted in any convenient location on the dash. It is held in place by means of screws through the surface of the U-shaped channel of the minibox.

A parts list for the Windshield-Wiper Controller is given in Table 17-1.

Table 17-1. Windshield-Wiper Controller Parts List

C1—50-μF, 15-Vdc electrolytic capacitor
D1—silicon diode, type 1N5060 or equiv
Q1—2N2646
R1—1-megohm potentiometer, linear taper (Mallory U54 or equiv)
R2—22,000-ohm, ½-W resistor
R3, R4—220-ohm, ½-W resistor
S1—miniature spst toggle
SCR1—silicon-controlled rectifier (G.E. GEMR-4, G.E. C20B, or
    equiv)
Misc—3¼″ × 2⅛″ × 1⅝″ aluminum minibox (Bud CU-2101A or
    equiv); 1¾″×2¾″ perf board; sleeving for diode D1; heavy utility
    wire for hookup to wiper switch; extra nut for R1 shaft; 1¾″ × 2¾″
    cardboard antishort shim; angle bracket for SCR1.

Ever wish you had a "live" 115-volt wall outlet on a camping trip? With the portable transistorized inverter shown in Fig. 18-1, you can have one. It gives you approximately 115-volts ac from a 12-volt car or camper battery so you can have a 115-volt wall outlet right on your dashboard. While the power capacity is limited to 100 watts, this is enough to run many small electrical conveniences you normally enjoy only at home, such as a stereo phonograph, radio receiver, tape recorder, electric shaver, lamps up to 100 watts or even a small tv set.

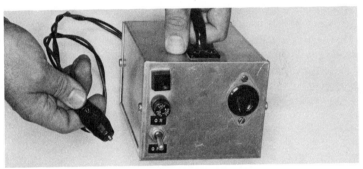

Fig. 18-1. Portable Ac Outlet.

The parts all fit in a 6-inch × 5-inch × 4-inch aluminum minibox with a carrying handle on top. This unit is small enough to store easily in a glove compartment. A flush-type receptacle is mounted on one end along with a pilot light and the on/off switch. You can plug directly into this outlet or use an extension cord if you want power at some other remote location, such as on a tailgate in a station wagon or inside a tent near the car.

The inverter is designed for use with 12-volt, negative-ground systems, which are the most common type used today. It is very important to observe correct polarity so that the power transistors will not be damaged. For quick, easy hookup, the input leads can be wired to a handy cigarette-lighter accessory plug available at auto-supply stores. The plug can then be inserted in the cigarette-lighter socket on the dashboard whenever you want ac power. When not in use, the inverter can be unplugged and stored away. For maximum efficiency, however, it is best to connect the positive lead of the inverter directly to the positive battery terminal, and the negative lead to a ground connection on the car.

In either case, use heavy, 12-gauge stranded wire since the inverter leads must be capable of handling a hefty current flow. If the wires are thin or the connections weak, the ac output will be reduced.

At the heart of the inverter is a Triad TY-75A step-up transformer. As shown in Fig. 18-2, the center tap of the transformer primary (black lead) goes to the positive terminal of the 12-volt supply. The two halves of the primary winding are connected to identical transistor circuits. The transistors conduct alternately, producing a current flow first in one half of the primary winding, then in the other half. This, in turn, creates a stepped-up alternating current in the secondary winding. Since the transistors turn on and off 60 times a second, the current in the secondary winding has a frequency of 60 cycles to match that of regular house current. The transistors should be high-power germanium pnp types, such as the ones specified in the Parts List (Table 18-1). If you consider other types, be sure they have the following minimum specifications: breakdown voltage of 36 volts, beta of 50, and power rating of 150 watts.

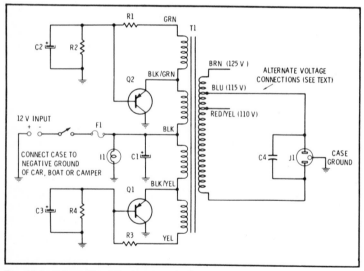

Fig. 18-2. Schematic of the inverter circuit.

How It Works

When switch S1 is closed, a current step is applied to the bases of both transistors Q1 and Q2 through the center-tap connection of

Table 18-1. Portable Ac Outlet Parts List

C1—250-μF, 25-Vdc electrolytic capacitor
C2, C3—2-μF, 50-Vdc electrolytic capacitor
C4—1-μF, 400-Vdc capacitor
R1, R3—5.1-ohm, 5-W resistor
R2, R4—200-ohm, 20-W resistor
T1—step-up transformer (Triad TY-75A or equiv)
I1—12-V indicator lamp assembly (Industrial Devices B3060D1, or equiv)
J1—single ac receptacle
Q1, Q2—high-power germanium pnp transistor (Motorola HEP G6009 or equiv); alternative transistors; 2N3637, 2N3638, or 2N3639
S1—heavy-duty spst toggle switch, 10-A minimum rating
F1—10-A 3AG fuse and fuseholder (Little fuse 342001A or equiv)
Case—6″ × 5″ × 4″ aluminum minibox (Bud CU-2107A or equiv)

T1 and biasing resistors R1, R2, R3, and R4. Either Q1 or Q2 will turn on faster than its mate (it does not matter which). As one transistor reaches saturation, current flow through that half of T1 connected to the emitter of the saturated transistor becomes steady. Energy stored in the field surrounding that half of the transformer primary winding is instantly restored to the circuit as the flux collapses back through T1. This induces a large pulse in the primary winding which is applied to both transistor bases. The pulse polarity drives the transistor that is on toward off, and it drives the transistor that is off toward on. When saturation is reached in the second transistor, the cycle repeats. Capacitors C1, C2, C3, and C4 filter out transients that might adversely affect the circuit operation. The output power is taken from the tapped secondary of transformer T1. In the event of a short on the secondary circuit, energy needed to sustain oscillation is drained from the primary circuit without damaging the transistors. Fuse F1 protects the primary circuit.

As shown in Fig. 18-3, the transformer is bolted to the end of the minibox but is placed on the bottom to carry its weight. The transistors are mounted on the outside at the back (Fig. 18-4) for good ventilation and cooling. Because these transistors handle large currents and get warm, it is also important to fasten them firmly to the minibox so that it acts as a good heatsink. When connecting the transistors, note the markings on the leads to determine which is the base and which is the emitter. Most large power transistors of the type used here are stamped with a "B" and "E" to indicate the base and emitter. The metal case itself serves as the collector. If the transistor is not marked, hold it so that the two leads are horizontal

Fig. 18-3. Internal view of the inverter.

and slightly above the centerline of the case. The lead on the left is the base, and the lead on the right the emitter.

You will find there are three output taps on the transformer, which offers you a choice of voltages. It is necessary to select the tap that gives the closest to the desired voltage. Start with the blue lead and check the output voltage with a voltmeter. If the voltage is low, disconnect the blue lead and try the brown one. If the voltage is high, use the red/yellow lead. If a voltmeter is not available, compare the intensity of a 100-watt light bulb first on regular house current, then on the inverter. The bulb should have about the same

Fig. 18-4. Location of power transistors on back of inverter.

brightness on either source. If it seems too dim, connect the higher-voltage tap; if too bright, connect the lower-voltage tap. As shown in Fig. 18-3, use lug-type terminal strips for making connections between components. Do not operate the inverter without a load plugged into the outlet.

19. OIL-LEVEL CHECKER

Very few motorists are aware of the fact that accurate oil-level readings are only possible after the engine has stopped for at least two hours, giving the oil a chance to drip back into the pan. The "idiot" light or the oil gauge in your car reads oil *pressure*, but you must still make sure that you have enough oil to maintain proper lubrication.

With the electronic Oil-Level Checker shown in Fig. 19-1, you can now be sure of proper oil level before you begin the day's driving. All you have to do before you start your engine is to press the test button on the programmed Oil-Level Checker, and logic circuits will read a remote sensor and tell you if the oil level is all right or if you need to add a quart. This is all done from *within* the car, without lifting the hood, or pulling the dipstick.

Fig. 19-1. Oil-Level Checker.

How It Works

The transistors and diodes combine to make up the *timing circuit,* the *voltage-sensitive switch* and the *logic circuit,* as shown in Fig. 19-2. The timing circuit controls the power applied to a small glass-encased thermistor that is attached to the car's dipstick, opposite the "add one quart" mark. This circuit assures that power is applied long enough to try to heat the thermistor, thereby attempting to change its resistance. Transistor Q1 and Q2 form a monostable multivibrator with circuit values selected to give a time constant of about 25 seconds.

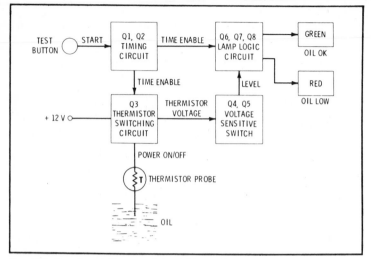

Fig. 19-2. Functional diagram of the Oil-Level Checker.

The schematic diagram for the oil-level checker circuit is shown in Fig. 19-3. When power switch S1 is in the on position, transistor Q3 is on and no voltage is available across thermistor T, since the circuit is shunted by Q3. This is the normal state of the circuit when power is applied. When the test switch is depressed, the base of transistor Q1 is pulsed with a positive voltage, which causes its collector voltage to drop to zero volts and charge the two parallel-connected 200-μF capacitors (C2 and C3). Transistor Q2, which is normally on, is turned off by the charge on the capacitors, and its collector voltage rises to 12 volts. Resistor R5, connected from the collector of Q2 to the base of Q1, feeds back sustaining voltage to maintain the monostable condition for a period of 25 seconds, as determined by the time constant of C2, C3, and R3. While the timing circuit is in operation, transistor Q3 is turned off and 12 volts is applied to the thermistor through resistor R7. Current flows through the thermistor for a period of 25 seconds.

The characteristics of the thermistor are such that, with current flowing through the device, its resistance will decrease as a result of the increase in its absolute temperature. The temperature of the thermistor is a function of the physical area of the thermistor. Thus, if we increase the effective thermal radiation area of the thermistor by placing it in contact with the cold oil in the crankcase, we have essentially a huge "heatsink" capable of absorbing the heat of the thermistor. The heatsink prevents the thermistor from

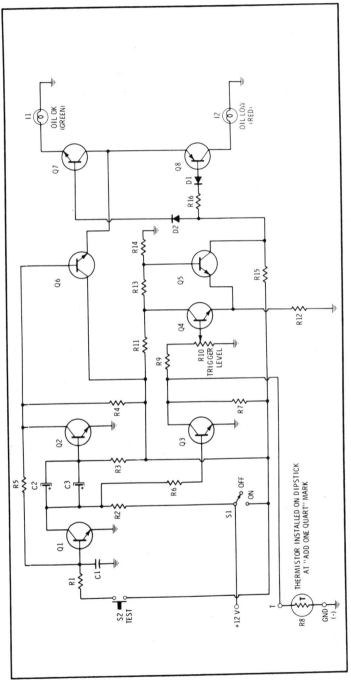

Fig. 19-3. Oil-Level Checker schematic diagram.

heating itself up and decreasing its resistance. However, if the thermistor is not in contact with oil, its temperature will increase rapidly due to the current flowing through it and, hence, its resistance will decrease sharply.

A voltage divider consisting of R9 and R10 is connected across the thermistor, and whatever voltage appears across the thermistor will also appear across the voltage divider. Potentiometer R10 permits a trigger level to be set for the Schmitt trigger consisting of Q4 and Q5. This circuit is used to detect the voltage level across the thermistor (which corresponds to its resistance), thereby detecting whether the thermistor is in contact with oil or is in free air. With the power switch on and the test switch depressed, transistor Q4 senses the relatively high voltage level present across the thermistor, which has not yet begun to heat. As a result, Q4 becomes conductive. This turns off transistor Q5, and a positive voltage is available at its collector, which is coupled through diode D2 to the base of Q7. Transistor Q6 and Q7 connected in series form an AND circuit. Transistor Q6 receives an enabling input from the collector of Q2, which becomes positive when the test switch is depressed. The green lamp lights up, since both Q6 and Q7 are conductive.

So long as the thermistor resistance does not decrease, the voltage level available across R9 and R10 remains steady and the green lamp will remain on. This condition can occur only when the thermistor is in contact with cold oil. If the thermistor is not in contact with the oil, its resistance will decrease as it heats itself up within the timing period, and the voltage at the base of Q4 will fall below the trigger level of the Schmitt trigger. Transistor Q4 will turn off and transistor Q5 will turn on, causing the voltage at its collector to essentially fall to zero volts and thus removing the enabling input to transistor Q7. The green lamp then turns off because Q7 is no longer conductive. Simultaneously, when transistor Q5 becomes conductive it turns on transistor Q8, which, with Q6, forms a complementary AND circuit. With transistor Q6 still enabled by a positive input from Q2, and Q8 now becoming conductive, the red lamp will illuminate, indicating that the oil is low.

The green lamp and red lamp can *never* be on *simultaneously*. The logic of the circuit is such that *either* the green lamp *or* the red lamp *must* be on, but never *both*.

Let us now discuss the significance of the circuit. Power switch S1 is arranged so that it primarily controls power to the major circuits. However, the 12-volt supply is always connected to the collector of Q1 through resistor R2. Thus, an accumulated charge is maintained on capacitors C2 and C3 to ensure continuous timing accuracy and prevent false triggering during the period of time when the unit is first turned on.

Calibration

With the thermistor in free air, connect a dc voltmeter across its terminals. Turn the switch off and connect the oil-level checker to the positive and negative terminals of the 12-volt supply. Let the unit sit for two minutes to accumulate the initial charge on capacitors C2 and C3 to prevent false triggering. Depress the test switch and observe the thermistor voltage on the voltmeter. The voltage reading decreases from its initial reading of about 6 volts to about 2 or 3 volts as the thermistor heats up. Let the unit cycle out, disregarding any indications of the lamps, and let the thermistor cool off for one minute. Then push the test button again and observe that the voltmeter reading decays after the button is released. As the voltage crosses the 3-volt level, adjust the potentiometer so that the green lamp, which should have been on, goes off and the red lamp turns on. This sets the switching level of the Schmitt trigger so that it flips back to the so-called normal conditions when the voltage decays to about one-half of what it was when power was actually applied to the thermistor. This is because after about 6 seconds of operation in free air, the thermistor resistance should have dropped from its initial value of 1000 ohms to about 500 ohms. Again, let the unit cycle out, wait one minute for the thermistor to cool, and depress the test button again. Note that the green lamp comes on and remains on for a couple of seconds. As the thermistor reheats and its resistance decreases, the voltage level drops to 3 volts, the green lamp goes off, and the red lamp comes on. The red lamp remains on until the timing cycle of 25 seconds is over, at which time the red lamp goes off. Wait one minute and then hold the thermistor between your fingers or immerse it in a glass of water. Again depress the test button. Note that the green lamp stays on as long as you provide additional heat-sinking action.

Assembly

Refer to Table 19-1 for a parts list. The unit is built on perf board, and the component layout is easily followed from the photographs. The lamps, switches, and terminal board are mounted on one-half of a minibox, as shown in Fig. 19-4. The circuit board is fastened by four 1-inch screws which go through 5/16-inch spacers to maintain proper spacing between the board and the minibox to

Table 19-1. Oil-Level Checker Parts List

C1—.1-μF, 100-Vdc capacitor
C2, C3—200-μF, 15-Vdc electrolytic capacitor
D1, D2—1-A, 50-V silicon diode, type 1N5060 or equiv
I1—12-V red indicator lamp (Industrial Devices B3060D1, Radio Shack 272-322, or equiv)
I2—12-V green indicator lamp (Industrial Devices B3060D5 or equiv)
Q1, Q2, Q3—npn silicon transistor, type 2N3414 or equiv (Radio Shack 276-2009)
Q4, Q5—npn silicon transistor, type 2N5172 or equiv (Radio Shack 276-2009)
Q6, Q7—npn silicon transistor, type D42C5 or equiv
Q8—pnp silicon transistor, type D43C5 or equiv
R1, R6—10,000-ohm, $\frac{1}{2}$-W ± 10% resistor
R2, R4, R7—1000-ohm, $\frac{1}{2}$-W ± 10% resistor
R3—62,000-ohm, $\frac{1}{2}$-W ± 10% resistor
R5—22,000-ohm, $\frac{1}{2}$-W ± 10% resistor
R8—glass probe thermistor, 1k at 25°C (Fenwal Electronics GB31M2—DO NOT SUBSTITUTE)
R9, R13, R15, R16—2700-ohm, $\frac{1}{2}$-W ± 10% resistor
R10—5000-ohm potentiometer (Mallory MTC53L4 or equiv)
R11—4700-ohm, $\frac{1}{2}$-W ± 10% resistor
R12—68-ohm, $\frac{1}{2}$-W ± 10% resistor
R14—6800-ohm, $\frac{1}{2}$-W ± 10% resistor
S1—spst subminiature toggle switch
S2—normally open push-button switch
Misc—5$\frac{1}{4}$″ × 3″ × 2$\frac{1}{8}$″ aluminum minibox (Bud CU-2106A, Radio Shack 270-238, or equiv), 4$\frac{1}{8}$″ × 2$\frac{3}{8}$″ perf board, hardware, wire, solder, etc.

Fig. 19-4. Internal view of the Oil-Level Checker.

prevent shorts. The oil dipstick in your car is modified by adding the thermistor so that the "bead" is just at the "add 1 quart" mark (see Fig. 19-5). The thermistor support is a small piece of dowel rod, shaped to be slightly narrower than the dipstick and grooved to keep the thermistor centered and away from the dipstick itself. The bead should also be about ¼-inch away from the support.

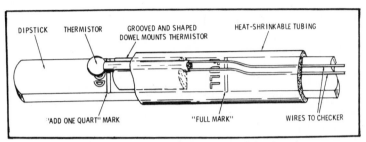

Fig. 19-5. Modified oil dipstick showing placement of the thermistor.

Carefully solder two insulated wires to the thermistor leads and slip short pieces of heat-shrinkable tubing over them. The thermistor and its support are held onto the dipstick with the heat-shrinkable tubing which is then shrunk by putting the whole assembly in an oven preheated to about 200°F. It may also be wise to reduce the width of the dipstick slightly along the length of the heat-shrinkable tubing to be sure that the width does not exceed the inside diameter of the dipstick hole.

Using the Oil-Level Checker

Connect the instrument's (+) and (−) terminals to a fused 12-volt line that does not go through the ignition switch, and connect the wires from the dipstick to the (T) and (−) terminals. Insert the dipstick to the usual depth.

Press the test button. Depending on your oil level, the green lamp will go on and stay on, or the red lamp will come on.

NOTE: The checker is meant to be used *before* start-up, while the oil level is at its highest and while the oil is at ambient temperature. Obviously, oil temperature will be high in a running engine, and this will prevent "self-heat" of the thermistor from being detectable, since ambient temperature is already high. (*No* device can measure oil level accurately in a running engine because the oil is not all in the reservoir pan but is circulating throughout the engine, where it should be!)

The road ahead is narrow and dark. As your car plunges into the gloom, its headlights splash twin pathways of illumination across the blacktop pavements. Beyond, the road, landscape, and sky merge in an inky, monochromatic blackness. Suddenly, a huge, jagged pothole flashes briefly into view. You swerve. *Too late!* The left front wheel drops. *Slam!* It strikes the opposite edge of the hole and rebounds upward, returning to the pavement with a hollow thud! You regain control, but your visibility now seems limited— one of the headlamps is not working. You flick the dimmer switch, but the low beams do not respond on that side either.

Is this condition serious? Safety experts say: *"Yes!"* Your car's sealed-beam headlamps are designed to provide good road illumination, but they also serve as visual "marker beacons" to the driver of an oncoming vehicle. Headlamp spacing and their height above ground help the approaching driver identify your vehicle as an automobile. The changing distance between the approaching two beams gives a fair measure of how far away your vehicle is and helps him to approximate the closing speed. Most important, however, your car's headlamps give the other driver an idea of the width of your car and where it is on the road with respect to his own vehicle. Take away *one* headlamp and all of these visual clues are lost to the approaching driver. Your car becomes a "one-eyed menace."

Detroit has put some changes into effect on late-model cars, however, that promise to eliminate this hazard. Through some special wiring provisions, carmakers have tied the parking lamps to the headlamp circuit so that both sets of lamps glow when the headlamps are turned on. If the headlamp fails, the parking lamps remain operative so that the oncoming driver still receives the visual clues he needs to pass by safely.

We both own vintage cars, neither of which has this desirable feature. We decided to add the feature to our cars, and while we were about it, a second idea presented itself. Like many other forgetful drivers who park and lock their cars and walk away oblivious to the headlights they have left blazing, both of us have suffered bruised egos and the inconvenience of dead batteries as a result of our forgetfulness. So, we have combined a parking-lights-on-with-headlights safety feature and an indicator to show that the headlights are on when the ignition is off. The result is the "Reminderlight" (Fig. 20-1).

Featuring just three major components (see the parts list in

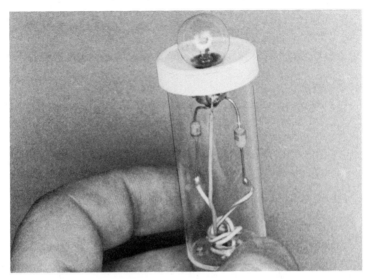

Fig. 20-1. The Reminderlight.

Table 20-1. Reminderlight Parts List

D1, D2—2-A, 50-piv silicon diode (RCA SK3081 or equiv)
I1—No. 257 self-flashing lamp
Misc—1″ dia × 2″ long plastic pill container, wire, solder

Table 20-1)—two silicon diodes and a self-flashing lamp—the Reminderlight assembles in only a few minutes. The two silicon diodes are soldered to the lamp and mounted inside a plastic pill container (usually available for the asking at any drugstore). The self-flashing lamp mounts in the soft plastic cap of the container, and the leads are soldered to the free end of the diodes and the shell of the lamps, as shown in Fig. 20-2. Ordinary 20-gauge hookup wire can be used for the leads. Refer to the schematic in Fig. 20-3 for the correct diode polarity.

An exit hole in the bottom end of the container permits the leads to be passed to the outside, where they are attached to the car's wiring. The diodes and leads are simply inserted into the snug-fitting container, and the plastic cap bearing the bulb is snapped into place, affording a neat, easy-to-install assembly.

How It Works

The schematic of the Reminderlight is given in Fig. 20-3. Only three leads need be connected to the car wiring. Lead 1 connects the

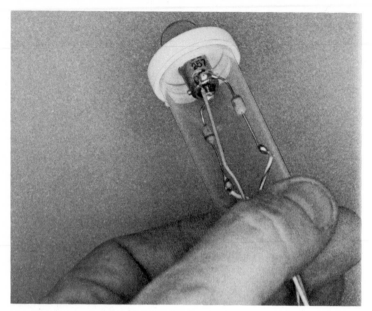

Fig. 20-2. Internal connections for the Reminderlight.

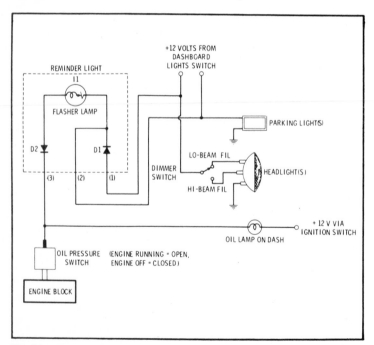

Fig. 20-3. Reminderlight schematic and external connections.

anode of diode D1 to the headlamp circuit, and lead 2 attaches to the parking-lamp circuit. Lead 3 connects to the wire that runs from the oil-pressure sensing switch (usually located on the side of the engine block) to the instrument panel OIL indicator. In most cars, you can make all the required connections without going under the hood, since all affected circuits are usually accessible under the dash. Because diode D1 allows current to flow in only one direction, current cannot flow through D1 to the headlamps when the parking lamps are on, but when the headlamp circuit is energized, D1 conducts so that power is also applied to the parking-lamp circuit. This simple arrangement yields the parking-lamp safety feature found in all late-model cars.

The reminder portion of the circuit comprises lamp I1 and diode D2. Lamp I1 is a special flasher lamp (available from the source specified in the Parts List). This lamp has a built-in bimetallic element which is in series with the lamp filament. When the filament is energized, it imparts heat to the bimetallic element causing it to bend. As it bends backward, it breaks the circuit to the filament and the lamp extinguishes. The bimetallic element cools and restores the power to the filament, and another cycle commences. The result is an attention-getting flashing action, bright enough to be seen in daylight.

To make the circuit operative, +12 volts must be present at the cathode of diode D1 (either from the headlamp or from the parking-lamp circuit), and a ground must be present on the cathode of diode D2. The oil-pressure sensor switch provides the necessary logic. When the engine is operating, oil pressure forces the contacts of this switch apart, breaking the circuit to ground. (That is why your oil indicator stays off as long as oil pressure is normal). But, when the engine is off, the sensor contacts close, effectively grounding the cathode of D2. This forms a ground path for current to flow through lamp I1, *if* you have shut off the engine but left the lights on. Of course, turning off the lights will remove power from the lamp.

Installation

The Reminderlight may be installed at any convenient point under the dash or on the steering column. You can fashion a strap or bracket to hold it in place or use plastic electrical tape to bind it to a convenient supporting member. Wire 1 may be connected to the lead that runs from the headlight switch to the floor-mounted dimmer switch. Connect it to the lead that goes to the movable switch arm so that the parking lights will be on, regardless of the dimmer switch position (high- or low-beam). Connect lead 2 to the wire that goes to the parking-lamp circuit. Lead 3 connects to the wire from your engine's oil-pressure sensor switch. In all cases, carefully skin the existing wire with a razor blade to expose the conductor. Then, wrap the lead from the Reminderlight tightly around the exposed conductor and insulate the connection with good-quality plastic electrical tape. Your best guide to wiring identification is the manufacturer's shop manual. If you do not have one, check with your local dealer's service department for the correct wire colors and hookup information.

Once you have put the Reminderlight into service, you will never have to pay any further attention to it. That is, not until it flashes a warning to save your battery by shutting off your lights, or until it averts a tragedy on the highway by keeping your car from becoming a "one-eyed menace."

21. VOLTMINDER FOR YOUR CAR

This easy-to-build instrument analyzes the condition of all automotive electrical circuits. It checks ground connections and voltage levels at the starter motor, starter switch, voltage regulator, generator or alternator, all lights, distributor, ignition primary circuit, and fuses.

The Voltminder shown in Fig. 21-1 is easily constructed. It is

Fig. 21-1. The Voltminder.

housed in a 3¾-inch × 6-inch × 2-inch plastic utility case whose metal cover is cut to fit a standard meter. The range-selector switch and binding posts are also conveniently located on the front panel.

A schematic and pictorial layout of the Voltminder are shown in Fig. 21-2. A parts list is given in Table 21-1.

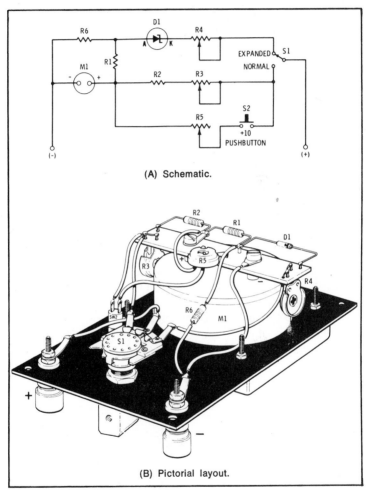

(A) Schematic.

(B) Pictorial layout.

Fig. 21-2. Schematic and pictorial layout of the Voltminder.

The normal range of this instrument measures direct (dc) voltages from 0 to 18 volts. The expanded range covers 11 through 16 volts over the full length of the scale. This yields a 3 to 1 scale-expansion factor, which permits precise observation of volt-

Table 21-1. Voltminder Parts List

D1—10-V, ½-W zener diode (Sylvania ECG5019 or equiv)
R1—4700-ohm, ½-W ± 10% resistor
R2—15,000-ohm, ½-W ± 10% resistor
R3, R4, R5—5000-ohm subminiature linear control (Mallory MTC53L1 or equiv)
R6—2200-ohm, ½-W ± 10% resistor
SW1—spdt wafer switch (Centralab 1460 or equiv)
SW2—spst subminiature push-button switch
M1—0-1 mA dc milliammeter (Lafayette 99F50874, Radio Shack 270-1752, or equiv)
Misc—6″ × 3¾″ × 2″ Bakelite case with aluminum panel (Radio Shack 270-627); perf board; cigarette-lighter adapter; wire; solder.

age levels within the selected range. Each voltage range is individually calibrated by means of miniature resistance trimmer controls. These controls (R3 and R4) are mounted, together with two fixed resistors and a diode, on a small perf board that is secured directly to the meter terminals. The trimmer controls are supported by their terminals, which are soldered directly to the leads on the other side of the perf board. Flea clips are used as tie points. Fig. 21-2B clearly shows the internal layout. Be sure that both binding posts are insulated from the metal front panel to prevent shorts. The Voltminder can be used either with test prods for servicing and troubleshooting or with an adapter that fits directly into the cigarette lighter for continuous monitoring of the battery voltage.

The indicating meter is a standard 0 to 1 milliampere movement with the scale calibration changed in order to adapt it for the Voltminder. A full-size new scale appears in Fig. 21-3. You can have a photostat made at any local print shop, or you can trace this illustration directly.

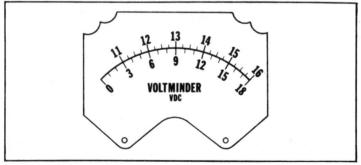

Fig. 21-3. Full-size calibrated scale for the Voltminder.

In order to achieve the expanded-scale characteristic, a zener diode is used. This semiconductor component has the ability to regulate voltage within precise limits. In this case we are using a 10-volt zener diode, which does not "break down" until at least 10 volts is applied. The zener diode prevents the meter from showing any indication at all until at least a 10-volt potential appears across the meter leads. Then, the diode allows current to pass and the meter begins to indicate. Thus, with the switch set to the expanded range, there will be no meter indication until more than 10-volts dc is applied to the Voltminder input terminals.

Calibration

It takes just a few minutes to calibrate both ranges of the Voltminder. This is done by adjusting the trimmer resistors (R3, R4, and R5) so that voltage readings correspond with those of another calibrated "standard" voltmeter.

Set the range switch to normal and apply a known dc voltage to the (+) and (−) terminals. This voltage could be from a car battery or from any variable low-voltage dc supply. Check the reading of the standard voltmeter. Adjust trimmer control R3 with a small screwdriver until the voltmeter indicates the same voltage as the standard voltmeter. Now switch to the expanded range and similarly adjust trimmer R4. Switch the Voltminder back and forth between the expanded and normal ranges. If the instrument is correctly calibrated, the readings will be identical for all voltages between 11 and 16 volts on both scales. Calibrate the ÷ 10 scale with the known voltage of a single flashlight battery. Connect the battery to the (+) and (−) terminals of both meters and adjust R5 until the standard meter and the Voltminder indicate the battery voltage (nominally 1.4 volts, which gives a reading of 14 on the Voltminder).

What Circuits Does the Voltminder Check?

Corresponding numbers in the typical electrical system of Fig. 21-4 show how the Voltminder tests every aspect of a 12-volt system. Since most of the tests check several components, there is considerable overlap, but this is ideal for isolating a problem. If you get a poor meter reading in all tests involving a given component, you know exactly where your trouble lies.

How to Use the Voltminder
Cranking Voltage Test

This test quickly determines whether or not sufficient voltage is being delivered to the ignition system while the engine is cranking. A normal reading confirms that the general condition of the battery, cables, starting system, and the circuit to the ignition system is satisfactory. An unsatisfactory reading may indicate that further testing or examination is required in this area. To perform the test:

1. Connect the (+) and (−) leads of the Voltminder to the ignition-coil primary and ground, as shown in Fig. 21-4.
2. Set the Voltminder switch to normal range.
3. Short the primary winding of the ignition coil with a clip-lead jumper or remove the lead from the ignition-coil tower so that the engine cannot start.
4. With the ignition switch on, crank the engine for five seconds and observe the Voltminder reading. The cranking voltage should not be less than 9.5 volts.

Results and Indications

Meter reads specified voltage or more; cranking speed normal and even—battery, starter, cables, switch, and ignition circuit to coil operating satisfactorily.

Meter reads less than specified voltage—weak battery, defective cables, poor connections, defective switch or starter, or defective ignition circuit to coil. (Check for defective ballast-resistor bypass in primary circuit for ignition coil.)

Cranking speed is below normal—excessive resistance in cables or starting motor. Excessive mechanical drag in engine (check oil grade).

Cranking speed is irregular—uneven compression, or defective starter or starter drive.

Charging Voltage Test

The charging voltage test provides a good overall indication of the voltage available to the entire electrical system. The voltage applied to the ignition system is an important factor to be considered when cases of distributor-point burning are encountered and when other electrical components are found to have subnormal operating lives.

In cases where subnormal charging-voltage readings are encountered, it is suggested that each component in the charging system be tested to determine exactly where the malfunction lies within the system.

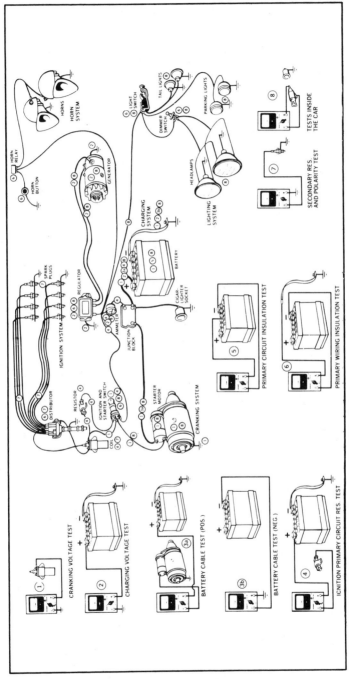

Fig. 21-4. Electrical hookup and tests using the Voltminder.

1. Set the Voltminder switch to the expanded position.
2. Observing the correct polarity, connect the (+) and (−) leads to the insulated post of the battery and to ground, as shown in Fig. 21-4, or to the battery terminal of the voltage regulator and to ground, depending on which is more accessible.
3. Operate the engine at a speed of 1500 to 2000 rpm.
4. Note the Voltminder reading after meter pointer stops climbing. The meter reading should be between 13.5 and 15.0 volts.

Results and Indications

Charging voltage is within the specified charging-voltage range for the vehicle being tested—charging system and voltage regulator operating satisfactorily.

Charging voltage is below specified range—defective generator (alternator) or drive system, defective or maladjusted voltage regulator, high resistance in charging circuit.

Charging voltage is above specified voltage range—defective or maladjusted voltage regulator, high resistance in regulator ground circuit, or defective field circuit.

Battery Cable Tests

Defective or undersize battery cables, loose or corroded connections, or excessively long cables can easily be the cause of inefficient starting-system operation, improper charging-system operation, or malfunction in many parts of the vehicle's electrical system. Defective cables, defective connections or both result in excessive voltage drop during operation of the starting motor. The battery should be fully charged, and the starter current draw must be within normal limits when battery cable tests are conducted. (If any doubt exists concerning the condition of the battery or the starting motor, these components should be tested by your local mechanic with a battery/starter tester.)

Positive Battery Cable Test

The entire insulated portion of a cranking circuit may be tested in one operation, or each individual portion of the insulated circuit may be tested separately to pinpoint an indicated defect.

1. Set the Voltminder switch to the normal position. Remove the lead from the ignition-coil tower to prevent the engine from starting.

2. Connect the (+) lead from the Voltminder to the center of the positive battery post and the (−) lead to the input terminal of the starting motor, as shown in Fig. 21-4 (3a), observing proper polarity. (Meter will attempt to indicate battery voltage until the ignition switch is closed.)
3. Operate starting motor and observe the Voltminder reading. Generally, the meter reading should not exceed 0.5 volt on most vehicles.

Negative (Ground) Battery Cable Test
1. Set the Voltminder switch to the normal position.
2. Connect (+) lead from the Voltminder to a ground on the engine block and the (−) lead to the center of the negative post of the battery as shown in Fig. 21-4 (3b). Repeat Step 3 of the previous test.

Results and Indications

Meter readings are within specified limits—cables, connections, etc., in normal operating condition.

Meter readings exceed specified limits—defective cable, undersized cable, loose or corroded connections, defective starter solenoid contacts, starter motor drawing excessive current, etc.

When above-normal voltage-drop readings are obtained, retest each item and connection within that portion of the circuit to determine the exact location of the fault. Correct the fault by cleaning and tightening connections or by replacing cables or components, as necessary, and then retest.

Ignition Primary-Circuit Resistance Test*

An excessive voltage drop in the ignition primary circuit between the battery and the ignition coil can reduce the secondary output of the ignition coil to the extent that hard starting and poor performance can result. This test checks the low-voltage ignition (primary circuit) wiring, including the ballast resistor (if used).

NOTE: On some vehicles, a special type of resistance wire is built into the wiring harness to serve the same purpose as a separate ballast-resistor unit.

1. Turn the Voltminder selector switch to the normal position.

---

*Where the vehicle is equipped with a solid-state ignition system, special checks may be required.

2. Connect the Voltminder leads as follows (observe proper polarity) : from battery (+) terminal to the high side of the ballast resistor, for coils equipped with an external ballast resistor, as shown in Fig. 21-4 (4). (From battery (+) terminal to the high side of the coil primary, for coils not equipped with an external ballast resistor.)

3. Use a jumper lead to ground the primary terminal of the distributor (lead from low end of coil primary). Grounding the distributor primary terminal with the jumper lead makes it unnecessary to spot the engine so that the breaker points are closed, and also eliminates the possibility of false test readings due to reduced current caused by defective points, wiring, and connections in the distributor.

4. Be sure all lights and accessories are turned off.

5. Turn the ignition switch on (do not crank engine) and observe the meter. The Voltminder should not read more than 0.5 volt.

6. Test the ignition switch by turning it off and on several times. The Voltminder should read the same each time the switch is turned on.

7. Test all ignition primary wires for tightness. Move them about and note any change in the meter reading with the ignition switch on.

Results and Indications

Meter is indicating within specified limits—connections, wiring, switch contacts, etc., in satisfactory condition.

Meter readings exceed the specified maximum—loose or corroded connections, undersized or faulty wiring, damaged or worn ignition switch contacts, etc.

If the meter readings exceed the specified maximum, isolate the point of high resistance by placing the test leads across each connection and wire, in turn. The reading across a connection should be zero. The reading across any one wire should be proportionate to its length, as compared to the length and allowable voltage drop of the engine circuit.

Primary Circuit (12-Volt System) Insulation Tests

Defective insulation in the primary circuit or at the battery can result in a constant loss of energy from the battery. Trouble of this nature is usually indicated by the fact that the battery becomes discharged if the weather is damp or if the vehicle has not been used for a day or two. Usually, leakage of this nature is so gradual that it is impossible to detect it on the charge indicator of the vehicle (if one is provided). These tests will reveal any major problem areas.

Electrolyte, dirt, moisture, or foreign material on the surface of the battery usually results in a continual battery discharge because this foreign material provides an electrical path between the battery terminals. Twelve-volt batteries are more susceptible to energy losses of this nature than six-volt batteries because of the higher voltage involved. It is always wise to keep the top of the battery as clean as possible to prevent formation of a leakage path. To check for leakage, proceed as follows:

1. Set the Voltminder switch to the normal position.
2. Connect the negative (−) lead from the Voltminder to the negative post of the vehicle's battery, as shown in Fig. 21-4 (5).
3. Move the positive (+) lead around the top surfaces of the battery, being careful not to touch the positive battery post.
4. Observe the meter for indications.

Any meter deflection indicates an energy loss due to dirt, moisture, or electrolyte on the external surfaces of the battery. When this condition exists, it is recommended that the battery be removed from the vehicle, thoroughly cleaned with a solution of baking soda or ammonia and water, and completely dried before reinstallation. It is also suggested that all dirt, moisture, corrosion, etc., be removed from the battery carrier. The cable connections should also be thoroughly cleaned. For dependable operation, make sure that the battery is in a state of full charge before reinstalling it in the vehicle.

Primary-Wiring Insulation Test

This test of primary-wiring insulation can detect leakage in the insulation that is too small to be located with an ohmmeter test. Considering the many accessories operated from the 12-volt line, it is not unlikely that leakage may occur, in time, in one or more circuits.

1. Disconnect the ground (−) battery cable from the battery post. Connect the Voltminder leads as shown in Fig. 21-4 (6).
2. Turn off all switches and close all doors to prevent operation of the courtesy light from the door switches. Disconnect the underhood or trunk light (if used).

3. Turn the Voltminder switch to the normal position.
4. If the vehicle is equipped with an electric clock, touch the battery cable to the battery post just prior to conducting the test, to rewind the clock, or leave the clock disconnected if your clock is the "continuous" type.
5. Note the Voltminder reading.

Results and Indications

Voltminder indicates zero—insulation in electrical primary circuits normal; no leakage exists.

Voltminder reads above zero—insulation leakage exists in one or more of the primary electrical circuits.

To locate leakage (an incomplete short) in the primary wiring circuits, remove the "hot" lead from each of the following components in the order in which they are listed. Retest the primary circuit for leaks after disconnecting each item.

Stoplight switch	Accessory switches
Courtesy light switches	Domelight switch
Horn relay and wiring	Regulator
Ignition switch	Heater switch
Light switch	Capacitors (rfi suppres-
Radio	sion and breakerpoint bypass)

NOTE: Capacitors may be located on the light switch, the regulator battery terminal, generator armature terminal, etc. Check your service manual for specific information.

Secondary Resistance Check/Polarity Test

Excessive ignition-coil secondary circuit resistance uses up energy that is needed to maintain good ignition under all conditions. It also severely reduces ignition-system reserve and consequently results in poor performance under severe operating conditions. Incorrect secondary-system polarity can result in up to 40 percent more voltage being required to fire the spark plugs, causing misfire and erratic engine operation. Polarity error can be easily corrected by interchanging the two primary-circuit connections at the ignition coil. Test as follows:

1. To prevent the generator from operating during the test, disconnect the lead from the field terminal of the generator.
2. Start the engine and adjust the speed to 1500 rpm.
3. Set the Voltminder switch to normal position. (Do not use

the expanded scale.) *CAUTION: Beware of high voltage at spark plugs: use an insulated probe on the negative lead.*

4. Connect (+) lead to engine ground, and contact the (−) test lead to each spark-plug-wire terminal, in turn, as shown in Fig. 21-4 (7). *Do not remove hv cable from the spark plug.* The meter should deflect upscale, indicating low resistance and correct spark polarity. All readings should be the same.

Results and Indications

Readings are upscale and essentially equal—secondary circuit in normal condition.

All readings are very low—corroded coil tower terminal, poorly connected or broken coil wire, center distributor-cap electrode burned, burned rotor tip, open secondary in coil.

One or more readings are lower than average—broken or poorly connected spark-plug wires, burned or corroded distributor-cap terminals, gouged electrodes inside the distributor cap.

Readings are higher than average at two or more spark plugs—crossfire occurring in the distributor cap or between the spark-plug cables concerned.

Meter reads downscale—polarity error, reverse the primary-circuit connections at ignition coil.

Battery Tests Inside the Car

With no current drain (all switches off), a good 12-volt battery should read between 12.2 and 12.8 volts, and a 6-volt battery should read between 6.1 and 6.4 volts. Voltages less than these indicate a weak battery.

It is normal for the battery voltage to drop somewhat under a heavy current drain—for example, when the starter motor is cranking the engine. The voltage of a good 12-volt battery may drop to about 9.5 volts, and a 6-volt battery may drop to about 4.75 volts under a 150- to 300-ampere starting load. A voltage drop of more than this indicates either a worn-out battery, poor cables or contacts, or a battery of insufficient capacity.

With the engine at a fast idle (about 1500 rpm) and all lights and accessories turned off, the charging voltage at the battery should be between 13.2 and 15.2 volts for a 12-volt battery and between 6.6 and 7.6 volts for a 6-volt battery.

Charging voltages lower than 6.6 volts in 6-volt systems or lower than 13.2 volts in 12-volt systems indicate either a faulty generator or alternator, a faulty or improperly adjusted voltage regulator, or a worn-out battery. Higher than specified voltages (generally 7.6 volts in 6-volt systems and 15.2 volts in 12-volt systems) indicate a faulty or improperly adjusted voltage regulator, poor battery-cable connections, or a defective battery. These values should be checked against the specifications for the electrical system being tested.

To make a Voltminder test of the battery from inside the car, proceed as follows:

1. Connect the cigarette-lighter adapter to the Voltminder leads (+ to tip; − to shell). Plug into the cigarette lighter to monitor the system voltage, as shown in Fig. 21-4 (8).
2. Check the battery condition from the front seat by observing the Voltminder (on the expanded scale).
   a. Engine off; note voltage.
   b. Turn on lights; note voltage.
   c. Crank engine; note voltage.
   d. Engine running; note voltage.

Results
   a. Engine-off voltage should be 12.0 volts.
   b. Lights-on voltage should be at least 11.8 volts (low beam).
   c. Cranking voltage should be 9.5 to 10.0 volts (at 70°F).
   d. Engine-running voltage should be at least 13.2 volts.

NOTE: The voltage readings given are for most makes and models of cars. Voltages may vary from car to car, based on battery age, condition, and temperature.

22. WINDSHIELD-WASHER FLUID WATCHER

Car windshields that are covered by mud spray and slush during winter months and after a summer rain constitute a significant driving hazard. Therefore, windshield washers are an indispensable aid in maintaining safe visibility on crowded highways.

We believe that an indicating system that monitors the washer reservoir to warn you to replenish the windshield-washer fluid before it goes dry is attractive "life insurance" to the driver. The Windshield-Washer Fluid Watcher shown in Fig. 22-1 is such a system.

The Circuit

Figure 22-2 shows a very simple circuit for a windshield-washer low-water detector. This circuit relies upon the conduction

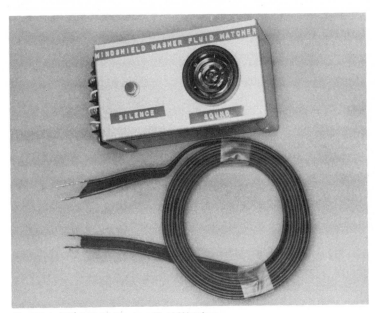

Fig. 22-1. Windshield-Washer Fluid Watcher.

of a minute current between two conductive probes suspended in the washer-fluid reservoir. Any detergent/antifreeze solution mixed with water makes it a better conductor of electricity. (Ordinarily, there are enough minerals present in most water supplies to guarantee that there will be some current conduction even when ordinary tap water is between the probes.) The "probes" are simply

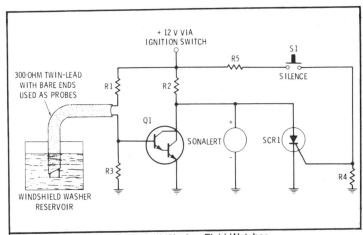

Fig. 22-2. Circuit for Windshield-Washer Fluid Watcher.

111

made from a tinned length of 300-ohm tv lead-in wire, with the exposed wire ends located approximately one to two inches from the bottom of the reservoir. One of the probes is connected to R1 and then to +12 volts via the ignition switch. The opposite probe connects directly to the base of a Darlington transistor to minimize the effects of temperature on transistor conduction. Ordinarily, the minute input current to the high-gain Darlington stage maintains the transistor in saturation. Collector current is limited at 60 milliamperes by resistor R2. The saturation voltage at the collector of Q1 (approximately 1 volt) is not sufficient to energize the Sonalert audible signal device. However, if the fluid level falls so that conduction does not take place between the probes, Q1 shuts off and its collector voltage goes more positive. This allows current to energize the Sonalert, which sounds a tone to tell you that the reservoir fluid level is low.

Silicon-controlled rectifier SCR1 and switch S1 gives you the ability to silence the Sonalert once its warning has been given. The SCR normally does not conduct until a positive pulse is applied to its gate. This is accomplished with switch S1. When its gate is pulsed, the SCR instantly turns on and latches (remains conducting). This deprives the Sonalert of operating power, since current through R2 is now shunted around the Sonalert, through the SCR to ground. The Sonalert thus remains effectively silenced until the SCR holding current is cut off by breaking the lead to the +12-volt supply.

Power is furnished through the ignition switch contacts, so that the circuit automatically resets every time the switch contacts are opened. However, if the water level is low the next time you start the engine, the Sonalert will sound again, until you silence it with a push of the button. With enough of these gentle reminders, you will take the hint and replenish the washer-fluid supply the next time you gas up!

Construction

All the components are installed in a 4-inch × 2¼-inch × 2¼-inch aluminum case, which can be mounted at any convenient location under the dash or on the steering column. A parts list is given in Table 22-1. The parts layout for the Windshield-Washer Fluid Watcher is shown in Fig. 22-3. (If you want the fluid watcher located out of sight, switch S1 can be located wherever you choose and the case can be bolted behind the dash.)

The "probe" that goes into the washer reservoir is a length of flat 300-ohm tv twin lead (two wires spaced about ½-inch apart in

Table 22-1. Windshield-Washer Fluid Watcher Parts List

Q1—Darlington transistor, type 2N5306 or equiv
R1—10,000-ohm, ½-W ± 10% resistor
R2—470-ohm, 1-W ± 10% resistor
R3—2.2-megohm, ½-W ± 10% resistor
R4—100-ohm, ½-W ± 10% resistor
R5—1000-ohm, ½-W ± 10% resistor
SCR1—silicon-controlled rectifier (Motorola HEP R1001, Radio Shack 276-1067, or equiv)
Sonalert—P. R. Mallory type SC628 or Radio Shack 273-060
S1—spst normally open push-button switch (Switchcraft type 101, Radio Shack 275-1547, or equiv)
Misc—4″ × 2¼″ × 2¼″ aluminum minibox (Premier PMC-1003, Radio Shack 270-236, or equiv); length of 300-ohm twin lead; perf board; wire; solder; etc.

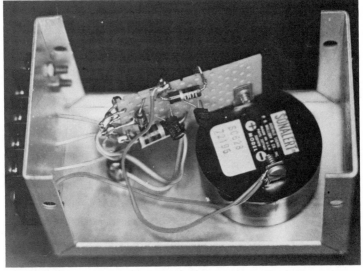

Fig. 22-3. Parts layout for the Windshield-Washer Fluid Watcher.

sturdy plastic insulation). The probe end is cut on a bias as shown in Fig. 22-4, so that one exposed wire is higher than the other. The exposed ends are tinned to prevent chemical reaction with the washer fluid. The flat ribbon wire is easily slid under the snap-on filler cap if the plastic neck of the reservoir is notched with a file. The wire is then fed back into the driving compartment through one of the firewall holes that are used for existing wiring.

It is important that the +12-volt supply line for the Windshield-Washer Fluid Watcher come from a point that is controlled by

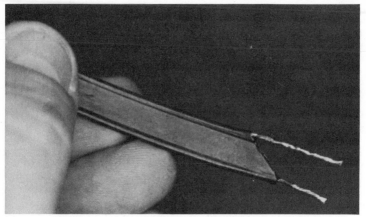
Fig. 22-4. "Probe" for Windshield-Washer Fluid Watcher.

the ignition switch. The radio and heater fan power sources are typical points that you can tie into at the fuse block. Be sure to check with a meter or test lamp to verify that the ignition switch actually removes voltage from the hookup point when the engine is switched off. Otherwise, the unit will remain permanently latched-on the first time you silence it!

23. COURTESY-LIGHT CONTROLLER

How often has this happened to you? You open the car door in the dark of night, the courtesy lights wink on to guide your way into the car, only to plunge you into darkness as you shut the door behind you! When you are burdened by packages or carrying a sleepy child, the dark interior of a car can be downright hostile! Leave the door open, you say? That's all right in summer, as long as rain isn't drenching you, but in the windy, chilly, drizzly real world, most of us would rather shut out the torments of nature and feel secure within the protective mantle that steel and glass afford. Part of that secure feeling is a well-lighted car interior.

Add this simple Courtesy-Light Controller (Fig. 23-1) to your car and you can feel at ease entering your vehicle in the dark. As you close the door, the lights remain on, giving you time to place your belongings, buckle your children into their seat belts, and find the ignition lock without scratching up the dash with an ill-directed key. After about one minute, the courtesy lights extinguish, but by this time, your eyes have become acclimated to the darkness and you are ready to go.

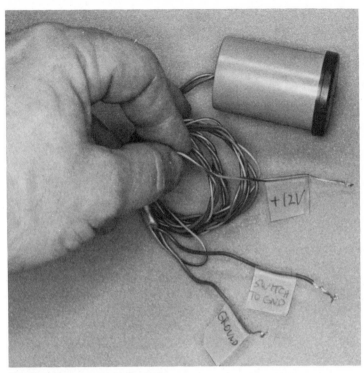

Fig. 23-1. Courtesy-Light Controller.

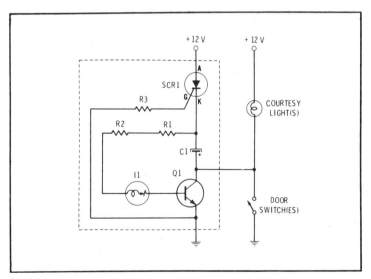

Fig. 23-2. Schematic diagram of the Courtesy-Light Controller.

The simple circuit that adds this touch of grace to your car is a one-transistor device that can be installed wherever you have room in your car and access to one of the door-activated courtesy-light switches. The circuit, shown in Fig. 23-2, relies upon the switching logic that is now used in most American cars: the switch that operates the courtesy light simply grounds the lamp return circuit when the door is opened. Closing the door breaks the ground path and extinguishes the light.

The courtesy-light controller circuit is connected across the switch—that is, between the lamp return lead and ground—so that when the switch opens (door closed), a path remains for current flow through the courtesy lights, keeping them on. The circuit is designed so that power to the lamps is shut off after a one-minute delay.

How It Works

In operation, npn transistor Q1 is the lamp switching device, with the collector connected to the lamp return lead and the emitter grounded. Silicon-controlled rectifier SCR1 and lamp I1 form a time-delay switching circuit that governs the conduction of Q1.

When the door has been closed for some time (normal condition), capacitor C1 charges to +12 volts, through R1, R2, lamp I1, the emitter-base junction of Q1, and the courtesy lights. However, Q1 is in the nonconducting state, as is SCR1. The gate of SCR1 is grounded through R3, and its cathode is raised above ground by the emitter-base junction of Q1 which functions as a diode.

When the door is opened, C1 is immediately discharged through the door switch. Instantly, the cathode of SCR1 is driven negative, effectively applying a positive-going pulse to the gate. The SCR immediately triggers, commencing current flow through R1, R2, lamp I1, and the emitter-base junction of Q1. This drives Q1 into saturation, supplying a return path for courtesy-lamp current flow when the door is closed again and switch S1 opens.

The SCR latches in the on state and, therefore, the courtesy lamps remain on. Meanwhile, the filament of lamp I1 is heating a built-in bimetallic arm. As time passes, the arm warps and eventually breaks the SCR cathode circuit. At that instant, SCR1 commutates and removes forward bias from transistor Q1. As a result, Q1 ceases to conduct, extinguishing the courtesy lights. As the bimetallic arm of lamp I1 cools, it restores the circuit, but the blocking SCR prevents current flow until it is once again triggered by a pulse from capacitor C1 as the door opens.

Construction

Refer to Fig. 23-2 and Table 23-1. Because transistor Q1 is operated as a saturated switch, it dissipates very little heat in the *on* state. Thus, there is no need for a heatsink when load currents are being switched within the free-air dissipating rating of the transistor. (Fortunately, this includes most courtesy-light circuit loads.) As a result, the controller can be housed in any convenient package. Our model fits nicely into a plastic can designed to hold a 35-mm film cartridge. The tightly fitted cap seals out dust and moisture. All parts can be wired point-to-point, as shown in Fig. 23-3, then tucked inside the can. Three wire leads exit from the can for connection to +12 volts, ground, and the door-switch circuit.

Table 23-1. Courtesy-Light Controller Parts List

C1—10-μF, 20-V electrolytic capacitor
I1—4.9-V, .3-A flasher lamp (General Electric type 408)
Q1—npn power transistor (Texas Instrument TIP3055 or equiv)
R1, R2—27-ohm, 2-W ± 10% resistor
R3—3900-ohm, ½-W ± 10% resistor
SCR1—silicon-controlled rectifier (Motorola HEP R1001, Radio Shack 276-1067, or equiv)
Misc—plastic film can with snap-on lid, wire, solder, etc.

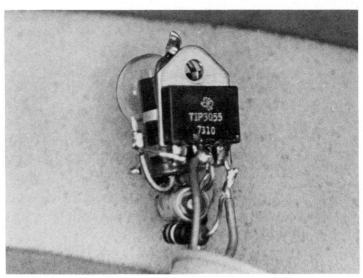

Fig. 23-3. Courtesy-Light Controller circuit construction.

Using the Courtesy-Light Controller

When all wiring has been correctly completed, open a door after a brief period with the doors closed and note that the courtesy lights come on. Now, close the door. The courtesy lights should remain on, then turn off after about a minute. This cycle should repeat each time the door is closed, then opened, and closed again. If you wish a shorter time delay, reduce the value of resistor R2 slightly.

24. QUICK-CHECK CIRCUIT TESTER

For a quick diagnosis of car wiring problems, this compact unit is a car-owner's best friend. Since it combines a polarity indicator with a continuity checker, it can answer questions like these: "Is a terminal connected to +12 volts or ground?" "Is the circuit from a given point completed to ground or another point?" "Is the wire, lamp, fuse, or switch 'open'?" "Has insulation been pierced, causing a wire that *should* show +12 volts to be accidentally grounded?"

All these, and more, helpful diagnostic checks can be performed on your car with this shirt-pocket-size instrument (Fig. 24-1). It's not fragile or bulky like a meter, and you don't need outside light to interpret its "readings." Just clip the ground lead to ground and touch the probe to the terminal or point being tested, as shown in Fig. 24-2. A red light-emitting diode (LED) flashes on if the terminal connects to +12 volts, or a green LED switches on to tell you that the point is grounded (or, that there is continuity between probe and clip lead).

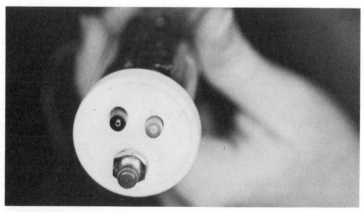

Fig. 24-1. Quick-Check Circuit Tester.

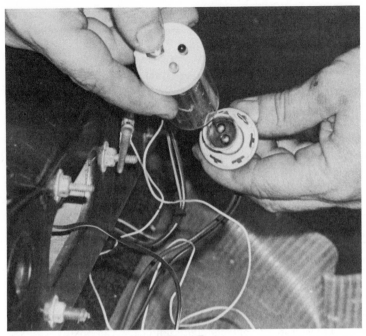

Fig. 24-2. Using the Quick-Check Circuit Tester.

The Circuit

The schematic of the Quick-Check Tester is shown in Fig. 24-3. The circuit relies on external voltage to sense the presence of +12 volts between the probe and clip, or relies on the internal battery to sense continuity between the probe and clip.

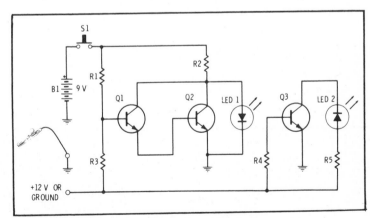

Fig. 24-3. Schematic of the Quick-Check Circuit Tester.

Transistors Q1 and Q2 comprise a high-gain Darlington amplifier. LED1 is connected between the collector and emitter of Q2. Ordinarily, when push-button switch S1 is depressed, Q1 and Q2 are instantly biased into conduction by current flow through R1. With transistor Q2 on, the battery voltage is dropped across R2 so that the only voltage appearing across green LED1 is the saturation voltage of Q2, which is insufficient to excite the LED.

Resistor R3 connects the base of the Q1-Q2 Darlington pair to the probe tip. The clip lead connects to the circuit common or ground, which leads back to the emitter of Q2. If the circuit between the probe and clip is open, LED1 remains off. If a positive potential is applied between the probe tip and clip, LED1 continues to be off since the positive input at the probe only serves to drive Q1-Q2 deeper into saturation. However, if the circuit is *completed* between the probe and clip, resistor R3 immediately "pulls down" the voltage at the base of Q1-Q2 so that insufficient current can flow to turn the Darlington pair on. Immediately, the collector voltage of Q2 rises, exciting LED1 into conduction. Similarly, if a sufficiently negative voltage appears between the probe and clip, Q1 and Q2 are biased and the LED1 illuminates.

Transistor Q3 controls operation of the red positive-voltage detector, LED2. It is responsive only to the presence of a voltage greater than +6 volts, applied between the probe and the ground clip. This voltage supplies forward bias to Q3 through resistor R4 and provides collector voltage through resistor R5 and LED 2. With the probe and clip open or shorted, no supply voltage is provided and LED2 is off. Furthermore, a negative voltage between the probe and clip reverse-biases LED2 and Q3 so that the LED remains off. Thus, only if the probe is applied to a wire or terminal that has a positive voltage with respect to ground (clip) does transistor Q3 conduct and turn on LED2. This checker is a reliable detector of the presence of voltage at a point, or of the presence of a path from the probe to the clip lead (ground, if you choose).

Resistors R1, R3, and R4 limit current flow in the base-emitter junctions of transistors Q1, Q2, and Q3 to a safe value. Resistors R2 and R5 limit the maximum current flow to LED1 and LED2 to about 20 milliamperes, a current that yields the best battery life and maximum brightness of the light-emitting diodes. The push-button switch saves the battery when the checker is not in use.

Construction

Refer to the parts list given in Table 24-1. Compactness and a

Table 24-1. Quick-Check Circuit Tester Parts List

B1—9-V battery (Mallory MN1604 or equiv)
LED1—green light-emitting diode (Lafayette 32-06349, Radio Shack 276-022, or equiv)
LED2—red light-emitting diode (Lafayette 32-06331, Radio Shack 276-041, or equiv)
Q1, Q2, Q3—npn silicon transistor, type 2N2926 or equiv
R1—100,000-ohm, ½-W ± 10% resistor
R2, R5—150-ohm, ½-W ± 10% resistor
R3, R4—15,000-ohm, ½-W ± 10% resistor
S1—miniature normally open push-button switch (Radio Shack 275-1547)
Misc—2⅞" × 1¼" diameter plastic pill bottle; 2½" × ⅞" perf board; battery snap terminals; ground lead with alligator clip; wire; solder; etc.

package that is small enough to fit into your hand are features of the checker's design. But, of course, you can vary the packaging to suit your needs and the availability of parts.

Our version is packaged in a 1¼-inch diameter × 2⅞-inch long plastic pill bottle with a snap-on top. The empty bottle easily accommodates a 9-volt battery, the two LEDs, transistors, resistors, and switch. The probe tip is a 1¼-inch finishing nail. (If you want an insulation-piercing probe, sharpen the tip to a fine point on a bench grinder.) The components nestle snugly inside the bottle. The LEDs and switch mount on the snap-on cap. Figures 24-4 and 24-5 show the parts layout in detail.

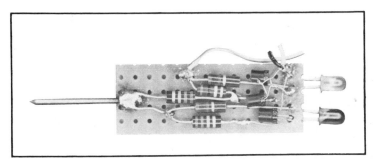

Fig. 24-4. Component side of circuit board.

Using the Checker

Clip the ground-lead wire to one end of the car circuit to be checked (ground to chassis, for example) and use the probe to touch significant points in the rest of the circuit, pressing switch S1 as you

Fig. 24-5. Complete assembly removed from the bottle.

do so. Where there is a positive voltage, the red LED will light; where there is a ground, the green LED will light. Neither LED will light with an open circuit. A bit of practice will make you a whiz at checking out auto electrical circuits with this easy-to-use helper.

25. TURN-SIGNAL ALERT

You flick the turn-signal lever as you commence a lane change or shallow turn. The lever latches and the turn-signal lights blink obediently, flashing a warning to vehicles ahead and behind that you are about to make your move. It's a nice day for driving—the sun is shining brightly, the window is down and the radio is turned up. Result? The turn signal keeps blinking long after your turn, because the steering wheel did not turn far enough to unlatch the turn-signal lever and you neither saw the indicator light, nor heard the soft clicking of the flasher.

This seemingly innocuous occurrence can have some grisly consequences. It leaves drivers of other cars guessing about your intentions and may lead to an accident under certain conditions.

The simple Turn-Signal Alert shown in Fig. 25-1 won't let that happen to you. Flip the lever to signal a turn and the circuit within the Alert is activated. It waits about a minute, giving you ample time to complete your turn or lane change; then it sounds off with a tone that pulses at the turn-signal flash rate to let you know that the lights are still blinking. Of course, if the turn-signal lever has been moved back to neutral (either manually or automatically), the circuit in-

Fig. 25-1. The Turn-Signal Alert.

Fig. 25-2. The Sonalert.

stantly resets until the next time the signals are switched on.

The pulsing tone is produced by a Sonalert (Fig. 25-2), a solid-state sounder that unleashes a distinctive 2500-Hz note that is easily heard through the din of traffic or over the noise of "hard rock."

How It Works

The turn-signal circuit used in most present-day cars is shown in simplified form at the top of Fig. 25-3. The directional switch is basically a single-pole, double-throw switch, and its arm is supplied with 12 volts through a flasher. The flasher contains a bimetallic strip that closes with a fixed contact when the circuit is open and no current is flowing. Moving the switch to either position applies the load to the flasher consisting of either the left or right set of signal lamps. The rush of current through the load heats the bimetallic strip, and it snaps to an open but unstable state, breaking the circuit. As the strip cools, it snaps closed again, and the cycle repeats until the turn-signal switch is moved to the neutral position, breaking the circuit to the lamps.

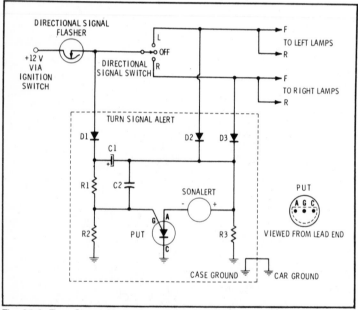

Fig. 25-3. Turn-Signal Alert schematic and external connections.

The Turn-Signal Alert is tied into the circuit at three points: diode D1 connects to the flasher; diodes D2 and D3 connect to the wires connected respectively to the left and right turn-signal lamps.

The circuit draws no current when the ignition switch is off and draws only a few microamperes in its sensing state when the ignition switch is on. In the latter state, +12 volts is applied through the closed flasher contacts across resistors R1 and R2, causing capacitor C1 to charge through R3. As long as the turn-signal switch is in the neutral (open-circuit) position, nothing happens. However, when that switch is moved to either the left or right signal position, +12 volts is applied through either diode D2 or D3, and the Sonalert to the anode of programmable unijunction transistor PUT1.

PUT1 is in the nonconducting state because its gate is more positive than its anode. In order to conduct, the gate must be at least 0.6 volt *below* the anode potential. At the instant that the turn-signal switch is closed, +12 volts is applied across resistor R3, and this potential appears in series with the charge stored on capacitor C1. As a result, the voltage appearing across R1 and R2 is now +24 volts, making the gate *very* positive with respect to the anode. Capacitor C2 helps to immunize the gate circuit against transients during this period.

However, as the flasher cycles, capacitor C1 commences to discharge through R1, R2, and R3 and to recharge *in the opposite polarity,* through diode D2 or D3. Gradually, the gate voltage falls until a point is reached where the capacitor recharge cycle has depressed the gate voltage below the anode voltage. At this point, the PUT triggers on and latches, supplying a ground path for the Sonalert. A pulsed tone now occurs each time the flasher circuit supplies +12 volts to the turn-signal lamps. If the turn-signal switch is moved to neutral, the PUT is deprived of holding current and the device blocks. Almost instantly, C1 recharges in the polarity to block the PUT and the circuit resets, awaiting closure of the turn-signal switch to commence a new timing period.

Construction

Refer to the parts list given in Table 25-1. The Turn-Signal Alert is housed in a miniature aluminum case, as shown in Fig. 25-4. The Sonalert mounts to the larger flat surface of one case half, and a four-terminal strip is secured to the top of that case half. All connections to the turn-signal circuit and ground are made at the screw terminals.

The PUT and the few circuit components are wired point to point, with the Sonalert terminals used as mechanical supports. If you wish, a perf board construction method can be used, but the simplicity of the circuit and the light weight of its components make this unnecessary. Be sure to observe correct polarity of diodes D1, D2, and D3, and electrolytic capacitor C1.

Table 25-1. Turn-Signal Alert Parts List

C1—25-μF, 25-Vdc electrolytic capacitor
C2—.01-μF, 200-Vdc disc or Mylar capacitor
D1, D2, D3—general-purpose silicon diode (Motorola HEP R0050, Radio Shack 276-1104, or equiv)
PUT1—programmable unijunction transistor (General Electric GE-X17 or equiv)
R1—470,000-ohm, ½-W ± 10% resistor
R2—2.2-megohm, ½-W ± 10% resistor
R3—2700-ohm, ½-W ± 10% resistor
Sonalert—P. R. Mallory type SC628 or Radio Shack 273-060
Misc—4″ × 2¼″ × 2¼″ aluminum minibox (Bud CU-2103A, Radio Shack 270-236, or equiv); 4-terminal screw-type terminal strip; wire; solder; etc.

Fig. 25-4. Internal view of the Turn-Signal Alert.

Installation and Checkout

Mount the assembled unit at any convenient point under the dash or along the sidewall where it will not be easily kicked and will not interfere with the accelerator and brake pedals. Run wire leads to the flasher and to the separate feed wires for the left and right turn-signal lamps. (If you have trouble finding the correct wires, check the color of the wires leading through the engine compartment to the front turn-signal lamps, or consult your car's shop manual.) Be sure to make a good ground connection to the car body.

When the hookup is complete, turn the ignition switch on and place the turn-signal lever in the left or right position. In about a minute, the Sonalert's tone will be heard, pulsing on and off with the turn-signal lamps. Flip the turn-signal switch to neutral and note that the Sonalert is silenced; then, move the switch to the opposite position for another timing cycle. The action should be the same.

If you find yourself trapped in traffic or at a long light, while signalling a turn, just move the turn-signal lever to neutral for an instant to silence and reset the Alert. The first time you hear its warning note when you thought the turn signals were off, we are sure you will agree that the few minutes spent in building the Turn-Signal Alert were well worth the effort.

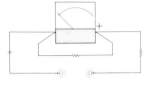

Fun and Games

Electronics is fun! Ask any hobbyist or experimenter. Besides giving you the fun of designing and building, electronics can add to your enjoyment of games, parties, picture taking and making, tapes, tv, and stereo. It can help you find lost metal objects, put rhythm into your music, "sniff out" radio-frequency energy so that you can operate ham and CB equipment with optimum results, enhance holiday lighting displays, and control electrical equipment just by whistling. It can give you the means to pit your skills against an opponent's in an electronic football game and can even help you make the "yes" and "no" decisions that confront you every day!

You can do all of these, and more, with the electronic projects described in this chapter. They're all designed for fun and game applications, and you can turn your imagination loose to use these designs in ways that enhance your leisure time.

Mechanical metronomes are known to almost every musician and student of music and have been in use for perhaps two centuries or more. In their simplest form they consist of a pivoted pendulum with one fixed and one movable weight. The position of the movable weight determines the number of oscillations or beats per minute when the pendulum is set in motion. More complicated versions employ a clockwork mechanism to drive the pendulum, and it is this version that is perhaps the best known.

But, technology moves on and electronics now offers the means to make a *better* metronome that never needs winding and

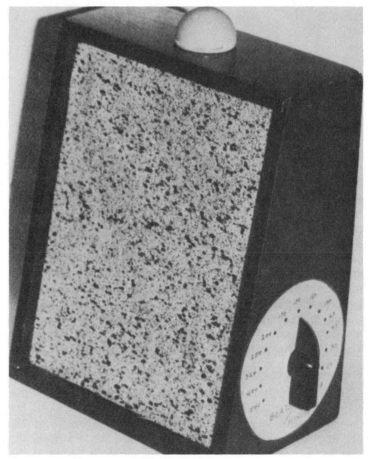

Fig. 26-1. Sight N' Sound Metronome.

Fig. 26-2. Side view showing power switch (SW1) and mode switch (SW2).

tirelessly taps out the rhythm to guide the student struggling to master any musical instrument.

The Sight N' Sound Metronome (Fig. 26-1) uses a completely solid-state circuit that is battery-powered for take-anywhere convenience. It features precise, adjustable control of beats per minute from a *largo* of 18 to a frenzied, high *presto* of 500 per minute. What's more, it produces these beats acoustically through a speaker and, if you wish, also flashes a light at the same rate. Alternatively, you may switch to sound only or to light pulses only for greater flexibility (see Fig. 26-2). Not limited to metronome use, the unit can be taken into the darkroom and preset to count out the seconds (audibly only, of course) in order to time photographic developing and printing steps. It can also be used to time party games. In its light-only mode, it makes a handy strobe source for determining speeds of rotating machinery or for producing attention-getting lighting effects for after-dark fun.

The Circuit

Figure 26-3 is a schematic showing the simple circuit of the Sight N' Sound Metronome. When SW1 is closed, C1 begins to charge through R1 and R2. Resistor R1 is the beat-per-minute (bpm) control. The lower its resistance setting, the faster C1 charges; the higher the resistance setting, the slower the charging rate. For any given setting of R1, C1 will eventually reach a voltage at which the emitter (E) of unijunction transistor Q1 is switched on. Instantly, Q1 conducts between E and B1, "dumping" the energy stored in C1 into an 8-ohm speaker. The brief current pulse through the speaker produces a distinct "plop" and, immediately, Q1 becomes nonconductive so that C1 again commences to charge.

If the mode switch, SW2, is in the sound-only position, the voltage pulse appearing across the speaker winding is blocked from lamp drivers Q2, Q3, and Q4 by reverse-biased diodes D1 and D2. Advancing SW2 to the sound-and-sight position switches D1 and D2 into the circuit. This means that the discharge pulse of the unijunction is applied to both the speaker (through D1) and the low-

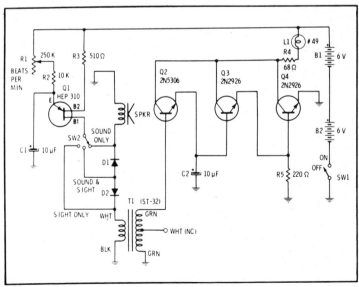

Fig. 26-3. Schematic diagram of Sight N' Sound Metronome.

impedance winding of a step-up transformer, T1 (through D2). Hence, as the pulse is heard in the speaker, it also is inducing a higher-voltage pulse across the base and emitter of Darlington amplifier Q2, Q3, and Q4. This cascaded-emitter follower circuit has exceptionally high current gain. Normally, the stages are non-conducting, so power cannot reach lamp L1, but as the brief pulse appearing across the T2 secondary drives Q2 into conduction, the extra gain of Q3 and Q4 are sufficient to briefly switch L1 on, then off, as the pulse wave passes. Capacitor C2 "stretches" the pulse slightly to overcome the thermal inertia of the lamp, so that a bright flash occurs. Limiting resistor R4 protects the 2-volt lamp against burnout at high flash rates.

In the sight only position of SW2, D1 and D2 block the discharge pulse of C1-Q1 from the speaker and only the lamp driver circuit (Q2, Q3, and Q4) is excited.

Construction

The Sight N' Sound Metronome (Fig. 26-4) is noncritical with respect to parts layout and can be scaled up or down to meet the size of just about any speaker housing you have handy. All electronic components other than the bpm control (R1), selector switch (SW2), power switch, speaker, and light are wired onto a piece of perf board. Clips can be used as tie points if you wish, or parts can be wired directly with no difference in performance.

In our unit, 6-volt lantern batteries were chosen as the power source, more for physical convenience than for current capacity. If you wish a more compact unit, AA size penlight batteries can be used in holders available from catalog houses. Merely wire the holders so that the resulting terminal voltage of the battery is 12 volts.

The 2-volt lamp, L1, can be mounted on the unit, as in our model, or housed separately in another unit connected by wires to the main unit. (A cylindrical plastic pill container, available at any drug store, is a good choice.) If you wish, an open-circuit jack can be mounted on the main unit so that a separate-unit lamp can be plugged in. Possibilities are limited only by your imagination.

If you elect to place the lamp in the speaker enclosure, you may wish to cover it with a light-diffusing dome for a more pleasing visual effect (see Fig. 26-5). A Ping-Pong ball, sawed in half, makes a very good choice. Simply mount the lamp and glue one of the ball halves to the outside of the enclosure.

All wiring is point to point, and no special lead dress is required

Fig. 26-4. Rear view showing location of the components in side of the case.

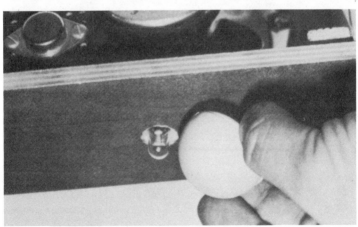

Fig. 26-5. Close-up of L1 and its translucent cover.

(Figs. 26-6 and 26-7). Soldering leads to batteries is permissible in this application, since the circuit current drain is so low that you'll achieve years of "shelf life" before replacement is required.

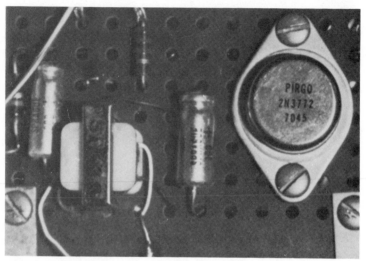

Fig. 26-6. Component board, showing layout of components.

Fig. 26-7. Reverse side of component board showing method of connecting leads together.

Calibration

When you have finished the construction and checked the wiring thoroughly, mount a circular white disc of paper or cardboard under the mounting nut that holds R1 in place. Affix a pointer knob to the shaft of R1 and set it to the maximum resistance position. Switch the power on and you should hear a periodic "plop" from the speaker, indicating that the circuit is working.

You can use another metronome for calibration, or you can calibrate by counting pulses in a one-minute period. For the higher repetition-rate range, you may want to use an audio oscillator or electronic frequency counter to check the exact rate. Scribe a pencil mark on the disc at each point you want to find again and write the repetition rate above it. This way, you'll be able to find a particular bpm setting without guesswork.

Using the Sight N' Sound Metronome

All beats produced by the unit are equally long and equally loud, regardless of the rate you choose. Since the filament has less time to cool between excitatory pulses at high rate settings, the flash rate of the unit is affected somewhat by thermal inertia of the lamp above 200 bpm. Nevertheless, for most musical rhythms, the unit will acquit itself admirably.

For solo use, place the unit where its beats will be audible to you and select a setting that matches the piece you are practicing. The audio pulse output is adequate for use atop a piano or organ in normal circumstances. For group practice, you may want to use both the sight and sound features of the unit, placing it where it can be seen, as well as heard, by all members of your ensemble.

In the photo lab, douse the sight feature of your unit and preset the bpm control for a one-second rhythm. This way, you can use your ears to track the progress of developing and printing your own pictures through each critical step.

For general experimentation, you will want to try "syncing" the pulsed light output of the unit to the operation of various machines. Performance is best with slower devices because the light output at higher repetition rates tends to be nearly continuous as a result of thermal inertia in the incandescent filament of lamp L1. Nevertheless, for "stop-action" studies of rotating parts, you will find the unit to be a worthwhile performer.

Table 26-1 contains a list of the components required for this project, with some suggested part numbers.

Table 26-1. Metronome Parts List

R1—potentiometer, 250k (Mallory Bridgetrol No. U44 or equiv)
R2—10k ± 20%, ½-W resistor
R3—510-ohm ± 20%, ½-W resistor
R4—68-ohm ± 20%, ½-W resistor
R5—220-ohm ± 20%, ½-W resistor
C1, C2—10-μF, 25-V capacitor (Sprague Type TE1204, Radio Shack 272-1013, or equiv)
B1, B2—6-volt lantern battery
SW1—spst toggle switch
SW2—sptt rotary switch (Centralab Type 1461 or equiv)
T1—transformer, 1k primary, 8-ohm secondary (Philmore Type ST-32, Radio Shack 273-1380, or equiv)
L1—No. 49 lamp (2-V, 0.06-A)
SPKR—8″ 8-ohm speaker
Q1—transistor (unijunction), HEP 310 or equiv
Q2—transistor (npn), 2N5306, HEP S9100, or equiv
Q3, Q4—transistor (npn), 2N2926, HEP 52, or equiv
D1, D2—general-purpose diodes (50-piv), HEP 154, or equiv
Misc—speaker coil, Ping-Pong ball, bayonette socket of L1, perf board, knobs

27. SHIMMERLITE

About to decorate that Christmas tree? If you have a couple of hours to spare, here's a simple project to really spruce up the old spruce. Just plug your light string into this Shimmerlite box (Fig. 27-1) and your tree lights gain an added strobe—or shimmer— effect. They pulse at any rate you wish, from a medium glow to a bright strobelike flash. Blinking lights? Add Shimmerlite to these, and they'll not only blink but also have added shimmer as well. You can use Shimmerlite to control outside lights, too.

Fig. 27-1. Shimmerlite in its finished state.

The Circuit

The Shimmerlite circuit (Fig. 27-2) ordinarily uses one-half of the ac cycle from your house current to power the lights, but as the timing-circuit capacitor charges, the voltage on the gate of the SCR builds. When it reaches the firing point, the SCR conducts, allowing the other half of the ac cycle to pass to the lights. The result: a quick shot of normal house current, giving the "strobe" effect.

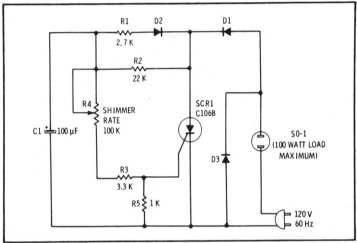

Fig. 27-2. Schematic diagram of the Shimmerlite.

Construction

You can easily fit the components for the Shimmerlite into a 1½-inch × 2-inch × 2¾-inch aluminum box (see Fig. 27-3). All parts, except the line socket and variable control, R4, are mounted on a piece of perforated circuit board. One-inch spacers support the board inside the box.

The SCR is mounted on a heatsink (made from scrap metal) to conduct heat away from the device. As Fig. 27-3 shows, the mounting tab makes the connection for the anode pin (pin and tab are connected internally in the SCR). *Because of this, be sure the heatsink does not touch the metal spacers or the box, once the cover is in place.* Component leads are soldered directly together and no board pins are necessary. Use 20-gauge hookup wire for the interconnections and keep them as short and neat as possible. For a more dramatic effect, you can build two or more Shimmerlites, connecting each to a separate string of lights. Each will flash at different rates and at different times.

Fig. 27-3. Shimmerlite with the cover removed.

If you wish to use Shimmerlite on an outside set of lights and cannot keep the box indoors, build the circuit in a water-tight container and do not leave the socket and control potentiometer exposed. Water can play havoc with circuitry—not to mention causing shocks.

One of the best features of this device is its small size. When it comes time to pack up the decorations until next year, Shimmerlite will easily fit in with your other accessories.

Table 27-1 contains a list of the parts required to build the Shimmerlite.

Table 27-1. Shimmerlite Parts List

```
C—100-μF, 50-V electrolytic capacitor
R1—2700-ohm, ½-W resistor
R2—22,000-ohm, ½-W resistor
R3—3300-ohm, ½-W resistor
R4—100,000-ohm potentiometer
R5—1000-ohm, ½-W resistor
D1, 2, 3—diode, Motorola HEP R0053, Radio Shack
   276-1104, or equiv
SCR4—silicon-controlled rectifier, GE C106B1, Motorola
   HEP R1221, Radio Shack 276-1067, or equiv
SO-1—line socket
Misc—line cord, case, perf board
```

28. FLIPPER

Two inexpensive ICs and a small handful of parts will give you an electronic "decision-maker" that says "yes" or "no" like a computer (Fig. 28-1). These parts provide *logic, memory,* and *readout*— the major ingredients found in every computer. The dual LED sends forth alternate rapid-fire bursts of red and green light that gradually slow down and finally stop—on either red or green. The decision is made! Is it "yes" or "no," do you "go" or "stop," is it "on" or "off," have you got "heads" or "tails," do you "win" or "lose"? If you are a

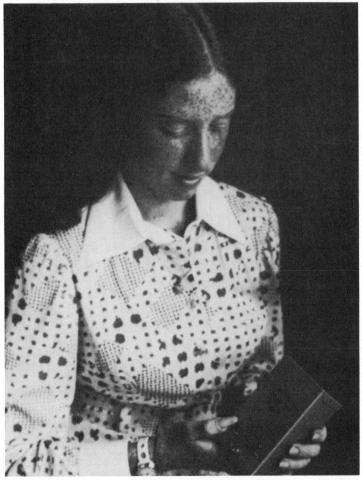

Fig. 28-1. Debbie Graf uses the Flipper to answer a question.

real sentimentalist, you'll know if she "loves you" or "loves you not." Flipper provides the answer to any question reducible to logic states.

WHAT IT DOES

Flip the power switch on and the LED starts to flash. Red/green/red/green/red/green—fast, but not so fast that the eye can't follow it. After a while, the flashing rate will gradually slow down until only one color stays on. (When you first turn Flipper on, it takes about a minute for the flashing to stop. Press the stop button, Fig. 28-2, for a few seconds if you don't want to wait.) Now we are ready to play.

Press the run button (Fig. 28-2) for a few seconds. The colors begin to flash—slowly at first, then faster and faster; the longer you hold the button down, the faster they flash. Let go, and the blinking begins to slow down and soon stops on red or green. *The longer you press the run button, the longer the flashing time.* The Flipper remembers! You can press the stop button at any time while the lights are blinking and they will instantly stop on red or green. Game possibilities for the Flipper are endless.

How It Works

One of the two inexpensive ICs used is an SN7405A hex inverter (Fig. 28-3). This means that within a tiny 14-pin package there are *six* complete amplifier/inverter circuits. Here's how they are used.

Three of the six circuits, together with only *one* capacitor, form a low-frequency oscillator whose output —a train of pulses—is fed through the fourth inverter to the input of an SN7470A flip-flop.

A flip-flop circuit works like it sounds. It's quite similar to a single-pole double-throw toggle switch in that it alternately turns on one of two circuits every time it is "flipped." That circuit stays "on" until the flip-flop is flipped once again. This marvelous electronic toggle switch has two output terminals called Q and $\overline{Q}$ and one input terminal called C. Each and every time a pulse (from the oscillator circuit) arrives at C it "flips" the circuit so that, alternately, there is an output signal at the Q and at the $\overline{Q}$ pins. Now, we take the remaining two inverter circuits, connect their inputs to Q and $\overline{Q}$ and their outputs to the bases of a differential amplifier. The collectors of Q2 and Q3, the npn transistors that make up the differential amplifier, are connected through resistor R7 and R8 pins of the 4½-volt supply voltage which is provided by three D cells.

Fig. 28-2. The Flipper.

The light-emitting diode (LED) is a special semiconductor that gives off visible light when it is forward biased. Depending on the chemistry of an LED, it gives off either red or green light. This particular LED contains two separate diodes and gives off *either* red *or* green light, depending on the polarity of the applied voltage. In our case, if terminal A is positive with respect to B, the lower diode

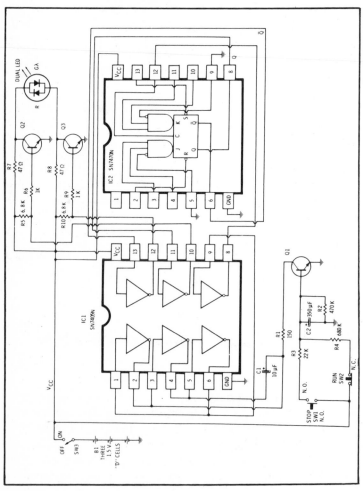

Fig. 28-3. The schematic diagram for the Flipper.

(green) will conduct. Reverse the polarity and the upper (red) diode will illuminate.

The LED is connected between the collectors of Q2 and Q3. Every time a positive voltage is applied to the base of either transistor, it turns "on." Since the bases get alternating positive inputs from Q and $\overline{Q}$ of the flip-flop, the transistors turn on alternately, which means *one collector or the other is always* at ground potential. In this way, either one-half of the LED or the other will conduct current through R7 and R8, respectively, and alternately, the LED will give off red or green light pulses.

The oscillator circuit is free running—going all the time—but if we connect one side of C1 to ground, the oscillation will stop. Make the connection *gradually* and the oscillator will *slow down* and *then* stop. This function is achieved by Q1, which acts as a slow-closing switch. When the unit is first turned on, capacitor C2, which is connected to the base of Q2, slowly charges through R4 until its voltage approaches the voltage level at which Q1 begins to turn on. This starts to slow down the oscillator and the LED flashing rate also slows. Soon Q1 is fully "on" and oscillation ceases. The flip-flop action stops and one-half of the LED *or* the other remains on.

To start oscillation again, push SW2. This opens the connection to +4.5 V, which allows capacitor C2 to slowly discharge through R2. Presently, the oscillator starts again—slowly at first, but soon it goes faster and faster. Release SW2, so that C2 can charge up again, and the process is repeated. The time it takes for the flashing to *stop* depends on how long *SW2 was depressed.* If you want an instant stop, press SW1. This charges C2 through R3—a much lower value resistor—so the stop-action is practically instantaneous.

Construction

A 5¼-inch × 3-inch × 2⅛-inch miniature box (Fig. 28-4) contains everything. Three D cells furnish power to the Flipper circuit, which is built on a 2-inch × 2½-inch perf board. The two ICs plug into two 14-pin dual in-line package (DIP) sockets that fit the holes in the board, making wiring quite easy. Three transistors, two capacitors, and three resistors are mounted on the same side as the ICs. The balance of the resistors fit nicely on the underside of the PC board (Fig. 28-5). Two small L-brackets are affixed with screws to the 2½-inch edge of the perf board. Each bracket is then held to the U-shaped portion of the miniature case by the push buttons. This provides a simple and secure mounting method for the circuit board.

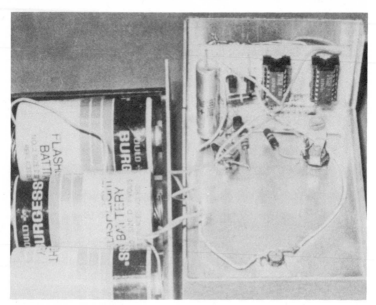

Fig. 28-4. The Flipper case opened, showing the location of the PC board and the battery pack.

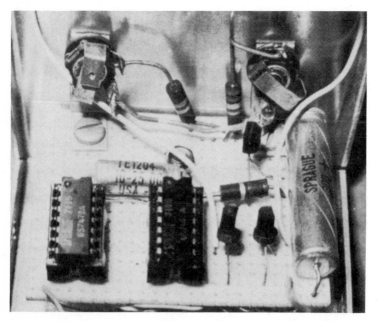

Fig. 28-5. PC board: showing component layout and mounting technique.

The on-off switch and the LED fit through two small holes, as shown.

The D cells fit into a three-cell battery holder that is held in place by four screws. The battery holder is located at the bottom half of the case so that there is enough room at the other end for the circuit board when the two halves of the case are fitted together. Labels and a few decorative stick-ons complete the job.

A parts list for Flipper is given in Table 28-1.

Table 28-1. Flipper Parts List

```
IC1—SN7405N hex inverter or equiv
IC2—SN7470N flip-flop or equiv
Q1—HEP730 transistor or equiv
Q2, Q3—2N4954 transistor (Radio Shack
    276-2009) or equiv
C1—10-μF, 25-V capacitor (Sprague TE 1204, Radio Shack
    272-1013, or equiv)
C2—350-μF, 16-V capacitor (Sprague TE 1166 or equiv)
R1—150-ohm resistor
R2—470k resistor
R3—22k resistor
R4—680k resistor          All resistors are
R5, R10—6.8k resistor        ± 10%, ½-W
R6, R9—1k resistor        composition type.
R7—43-ohm resistor
R8—47-ohm resistor
B1—three 1½-volt D cells
SW1—normally open push-button switch (Switchcraft
    "Tiny Switch" No. 961 or equiv)
SW2—normally closed push-button switch (Switchcraft
    "Tiny Switch" No. 962 or equiv)
SW3—miniature toggle switch
LED—red/green light-emitting diode (Monsanto MV5491, Radio
    Shack 276-035, or equiv)
Case—5¼" × 3" × 2⅛" minicase (Premier Metal PMC
    1006, Radio Shack 270-238, or equiv)
Misc—two 14-pin DIP sockets to fit 2" × 2½" perf board with
    1/16" grid; battery holder for three D cells (Calectro
    cat. No. F3-067 or equiv); wire; and solder
```

29. ELECTRONIC FOOTBALL GAME

It's your ball, fourth down, two yards to go for a first down, 18 yards to go for a touchdown. What do you do? Go for broke and try to pass for the score, or play it safe and stay on the ground, hoping to bull your way to a first down? This is the kind of excitement you can have with this Electronic Football Game (Fig. 29-1) based on a true, random-chance circuit. You select whatever play you want to try

Fig. 29-1. The Electronic Football Game.

from a chart—long pass, screen, end run, draw, quarterback sneak—and a light flashes on, telling you how many yards you gained or lost. There are chances for fumbles, interceptions, blocked kicks, long runbacks—all the things that happen in real football.

To make the game easy to build, we have included a scaled drawing of the playing field (Fig. 29-2) and the play-selection chart (Fig. 29-3) on these pages. Just reproduce them on a 15- × 15-inch square of ¼-inch hardboard to produce a neat, professional-looking game board. (If you wish, you may increase the size of the field and the chart.) The electronic parts are hidden in a recess under the board.

Although the game is based on chance, it is designed to offer the same statistical ratio of risks and rewards found in actual football plays. If you decide to try a long pass, for instance, you will find that the possible gains are bigger than they are for running plays but that there are also more chances the pass will fall incomplete or be intercepted. If you elect to stay on the ground, grinding out short yardage, the possible gains are smaller, but there are mathematically more chances to make them and less risk of losing the ball. You can thus work out your own strategy, just like the pros do.

148

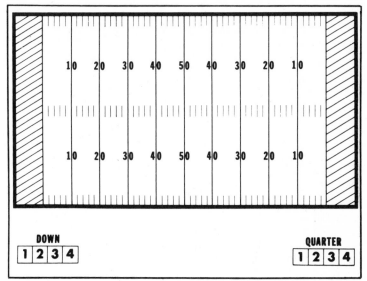

Fig. 29-2. Scaled-down diagram of playing field.

The play-selection chart shown in Fig. 29-4 has 16 neon lamps arranged in a row across the top. The circuit that controls them is called a ring counter because the lamps go rapidly on and off, one after another, around and around like a ring. They flicker on and off so fast—about 150 times a second—that they all appear to be on at once.

There are two push-button switches on the playing board, one for each player. Every play starts with each player pressing his button. This starts the lights, causing them to flicker simultaneously. The player with the ball then designates a play on the chart with his marker and releases his button. This breaks the circuit, and all bulbs go off except the one that was lighted at the instant the button was released. This bulb remains on, indicating the results of the play. If, for example, you chose a draw play and the second bulb remains on, you look down the chart and see that you made a 20-yard gain. If, on the other hand, the fifth bulb remains lighted, you were thrown for a two-yard loss.

You then start another play; each player presses his button and the lights go back on. It is possible for two bulbs to remain lighted after a button has been released. In this case, you replay the down.

Construction

When you lay out the game board, note that the small circles at the top of the play-selection chart represent holes for the 16 neon

149

Fig. 29-3. Scaled-down drawing of play-selection chart.

	○	○	○	○	○	○	○	○	○	○	○	○	○	○	○	○
KICKOFF	55	40	53	10	30	48	54	12	EZ	41	37	49	51	15	52	58
KICK RET	2	9	20	14	35	21	11	55	15	24	5	22	16	45	4	23
OFF TACKLE	-3	4	0	10	5	7	12	-2	6	0	2	15	3	-1	7	1
Q B SNEAK	1	1	0	2	-1	0	3	0	2	7	-1	2	0	4	2	1
FB CENTER	5	-1	2	0	7	3	1	6	3	-1	10	2	1	2	4	3
DRAW PLAY	1	20	17	0	-2	16	2	1	20	15	18	1	-2	2	0	2
OFF GUARD	4	-4	1	5	-1	2	0	7	3	3	10	-1	4	1	8	2
AROUND END	3	-1	-2	5	0	30	FUM	12	25	-5	45	15	3	20	-3	5
SCREEN PLAY	17	3	-7	0	20	-5	12	INC	5	45	3 INT	7	INC	0	4	-1
LONG PASS	5 INT	TD	INC	INC	25 PNF	INC	20 INT	INC	45	30	INC	25 INT	INC	35	30 PNF	INC
SHORT PASS	3	4	INC	7 PNF	5 INT	5	6	INC	20 INT	4	INC	7	7	10 PNF	INC	INC
INT RUNBACK	15	0	5	-4	20	7	FUM	TD	8	-3	0	5	40	4	10	FUM
PUNT	36	25	37	48	-8 BLK	35	47	38	45	-5 BLK	46	50	40	39	55	41
FIELD GOAL	12	23	20	29	BLK	27	18	28	40	14	BLK	35	15	BLK	38	30
FUMBLE	L	K	L	L	L	K	K	L	L	K	K	L	K	L	L	L
EX. POINT	G	G	NG	G	G	G	NG	G	G	NG	G	G	G	NG	G	G

EZ - End Zone FUM - Fumble INC - Incomplete INT - Intercepted BLK - Blocked
PNF - Pass Interference L - Lose K - Keep G - Good NG - No Good

150

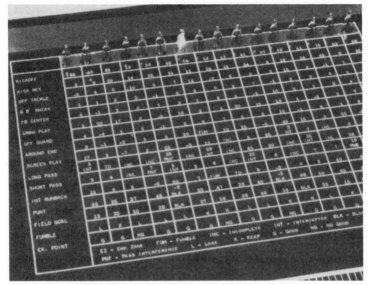

Fig. 29-4. Play-selection chart, with bulbs installed.

bulbs that protrude from under the board. Using a prick punch, mark these holes on the board and drill them out with a ¼-inch bit. Also drill holes for the push-button switches.

Paint the board green or black and use contrasting colors for the playing field and play-selection chart. To keep the play areas from becoming smudged and worn with age, you may want to tape a sheet of clear cellulose acetate over the top.

You can frame the board with strips of scrap wood, 2¼ inches wide, or you can use stock baseboard molding. The latter, turned upside down and mitered at the corners, gives the board an interesting tapered look.

All of the electronic parts are mounted on a piece of perforated phenolic board (Fig. 29-5) about 7 inches × 9 inches. Use flea clips to attach the parts. The circuit (Fig. 29-6) is actually simpler than it looks. Except for the power supply and relay, it consists of one basic grouping of parts which is repeated 16 times. The resistors and capacitors control the firing rate of the bulbs. The diodes short out the capacitors when either player's button is released. This reduces the voltage to the point where only the bulb lighted at that instant can remain on; the rest go off and do not have enough voltage to restart. This works because neon lamps characteristically require more starting-voltage than running-voltage.

Fig. 29-5. Phenolic board, showing component layout.

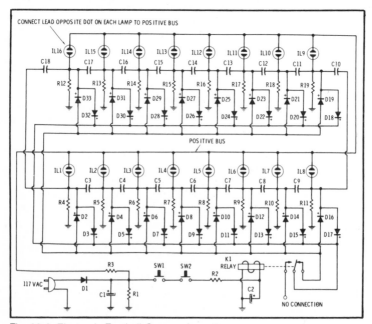

Fig. 29-6. Electronic Football Game schematic.

152

Be sure to observe correct polarity when installing the diodes or the bulbs will not light properly. A band or plus mark identifies the cathode end. The bulbs must also be connected in the proper way. You will find that one lead on each lamp is identified by a marker dot. Connect this lead to the positive bus wire. The bulbs are mounted on a separate cardboard strip (Fig. 29-7), so you will have 17 wires running from this strip to the main circuit board—one common bus and one individual lead from each of the 16 bulbs. Press the bulb tips gently up through the holes in the playing board and tape the strip to the underside.

Except for the neon lamps, all of the parts (see Table 29-1) are readily available. The neon lamps should be of a special type with closely matched characteristics so that they will fire uniformly and preserve a truly random pattern of operation. The lamps specified can be ordered from Inventive Electronics, Wykagyl Station, NY 10804. The price for 16 neon lamps is $10. A full-size, ready-to-use playing field and play-selection chart can also be ordered for $5 from the same company.

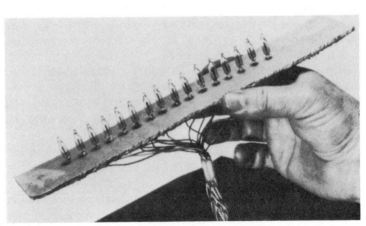

Fig. 29-7. Cardboard holder, with neon lamps installed.

Playing the Game

Except for a few necessary simplifications, the game is played according to standard football rules. Small radio-type pointer knobs or similar markers can be used to indicate the play selected on the chart, the position of the ball on the field, and the amount of yardage needed for a first down. Additional markers can be constructed to let you keep track of downs and quarters.

You start by flipping a coin to see who kicks off. The player

Table 29-1. Electronic Football Game Parts List

R1—330k, ½-W ± 20% resistor
R2—1k, ½-W ± 20% resistor
R3—100k, ½-W ± 20% resistor
R4 to R19—1.5-megohm, ½-W ± 10% resistors
C1, C2—20-μF, 150-Vdc electrolytic capacitors
C3 to C18—1-μF, 200-Vdc capacitors
D1 to D33—200-piv 1-A silicon rectifier diodes
IL1 to IL16—GE-5ABB neon indicator lamps (see text for information on where to buy.)
SW1, SW2—normally open push-button switches (can be radio type or simple, flush-mount doorbell type)
K1—normally open, single-pole, single-throw relay
Misc—line cord, perforated phenolic board, flea clips, wire, solder

winning the toss will receive and his opponent places the ball marked on his 40-yard line. The play marker is moved to the row labeled kickoff on the chart and the buttons are pressed. The light remaining on when the kicker releases his button indicates the kick's distance. If it went, say 48 yards, the ball marker is placed on the receiving team's 12-yard line. The play marker is now placed on *kick ret* and the process is repeated to see how many yards the receiver makes on the runback. (If the kick goes into the end zone, the receiving team takes the ball on its 20-yard line with no runback.)

At this point, the offensive player has a first down with four tries to go 10-yards and make another first down. He may use any scrimmage play shown on the chart, moving the ball ahead the indicated yardage each time. Minus yardage means a player has been thrown for a loss and the ball must be moved back the distance shown. Use sideline markers to indicate the position of the ball at each first down and the point the ball must reach for another first down. Another marker may be used to indicate the line of scrimmage.

A player may either kick on the fourth down or try for yardage. If he gambles and fails to make the first down, his opponent takes over the ball at that point. If he kicks, the play marker is moved to *punt* on the chart and the buttons are operated to determine the kick's distance. The opposing player then takes over the ball and uses the kick return play to determine how far he runs back the punt.

To score a touchdown on any play, the yardage indicated must be enough to reach or go beyond the goal line. The same holds true for a field goal attempt. If a field goal falls short, the ball is taken by the opposing player at his 20-yard line. If either a field goal or a punt is blocked, the opposing player takes possession at the line of scrimmage.

If the ball is fumbled, the play marker is moved to *fumble* on the chart and the buttons activated to see if the offensive team retains possession or not. If the result is "lose," the opposing player takes over at the line of scrimmage. If it is "keep," the offensive player loses the down but continues to play.

A pass can be complete with the yardage indicated, incomplete with no yardage, or intercepted. If it is intercepted, the ball is first advanced to the point of interception according to the yardage shown in the chart. Then the play marker is moved to *int runback* and the buttons actuated to determine how many yards the opposing player runs back the interception. If the intercepting player fumbles on the runback, the fumble play is used to tell whether he keeps possession at the point of interception or loses the ball. If there is pass interference, the offensive team gets a first down at the point of the interference which is indicated by the yardage shown on the chart.

As in regular football scoring, touchdowns count six points, field goals three, and extra points one. Conversion attempts are determined by using the extra point play to see if the kick is good or no good. You can, of course, vary the rules as you wish. You might decide, for instance, to permit two-point conversions.

Quarters can be determined by setting a kitchen timer (you might want to consider constructing and using the General-Purpose Timer covered in Project 36) to an agreed-on period or by limiting each player to a certain number of plays, say, 16 or 20. In this case, only actual plays from scrimmage count. Kickoffs, kick returns, interception runbacks, and conversion attempts are not included. At the start of the second half, the player who kicked off at the beginning of the game now receives the ball.

30. UNIVERSAL STROBE SLAVE

To paraphrase an expression, "two flashes are better than one" when it comes to taking pictures with a strobe or flashbulb. A single flash, directed from a camera at a relatively close subject, frequently plunges the background into unpleasant (and generally, undesirable) darkness. The disadvantage of a direct flash can be overcome by using bounce-flash techniques, but it then becomes quite difficult to calculate proper exposure. The solution is to use at least two strobe light sources: one "master" strobe at the camera location and a second one "slaved" to the first, but pointed from a different angle to soften the hard shadows. If desired, this second strobe can backlight the main subject, making it stand out from the gloomy background, or else it can be used to achieve uniform lighting throughout the entire picture area.

This easy-to-build Strobe Slave (Fig. 30-1) allows your camera-mounted flashbulb or strobe to fire a second strobe *without any interconnecting wires*, thus offering a lot of convenience and flexibility in lighting arrangements.

Fig. 30-1. The universal Strobe Slave.

The Strobe Slave has several unique features generally not found in the store-bought kind. It can be fitted with a long cord—up to 10 feet is practical—which allows you to face the Strobe Slave in the general direction of the triggering flash, guaranteeing sure-fire operation. The strobe unit can then be faced *anywhere*; it can be located so that light from the prime light sources only reaches it by reflection off the walls or ceiling.

The Strobe Slave has good sensitivity to the master flash and is not influenced by ambient light levels. It has even been used in bright sunlight. Its effective range depends on the reflectivity of the area between the slave and the master. Distances up to 50 feet have been realized indoors. Of course, several Strobe Slaves can be used for multiple-flash pictures without the need for any trailing wires between individual strobes.

The circuit and construction of the Strobe Slave are simple: one light-activated silicon-controlled rectifier. (LASCR) and one resistor, wired to a plug (Fig. 30-2). The schematic is shown in Fig. 30-3 and the method of assembly is entirely obvious from the photos.

If stiff wire is used, the assembly can be bent so that the

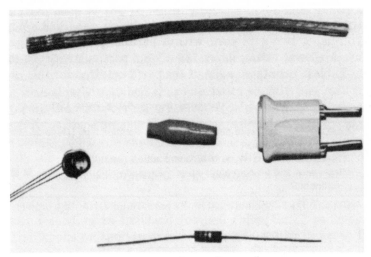

Fig. 30-2. Parts required to build the Strobe Slave.

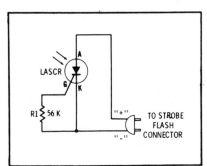

Fig. 30-3. Universal Strobe Slave schematic.

LASCR faces in any direction, giving optimum sensitivity to the master flash. The anode and cathode of the LASCR are connected to the flash contacts of the strobe, which is triggered by the strobe slave in such a way that the positive voltage from the remote strobe flash-unit contacts will go to the anode (A), and the common will go to the cathode (K). When the high-intensity flash of light from the master flash strikes the LASCR, a voltage pulse appears at the gate of the SCR sufficient to turn it on. At that instant, the anode-cathode circuit is completed and the strobe fires. If reversed, the unit will not work. (If you do not have a voltmeter, try the connection one way, and if the unit will not trigger, simply reverse connections.)

Many strobes are equipped with a receptacle somewhat similar to the ac receptacle in the power line except that one blade of the mating plug is generally narrower than the other. With some strobes, the narrower blade is positive, while in others, the larger blade is the positive side. Once you have determined the polarity of your strobe, simply mark the strobe slave plug so that it is always inserted correctly. (We used a regular line cord and simply ground one of the blades until it fits.)

If your strobe does not have this kind of receptacle, use a flash extension cord, sold at most camera stores. They have a female PC connector at one end and a male at the other. Simply cut off the connector that is not required and solder the wires into the proper terminals on the SCR. Whether you use a two-wire cord with a parallel plug or whether you use a special fitting depends on your particular strobe unit.

Table 30-1 contains a list of the parts required for this project.

Table 30-1. Universal Strobe Slave Parts List

LASCR—light-activated silicon controlled rectifier type
  (General Electric, Type L8-A or equiv)
R1—56,000-ohm, ½-W, ± 10% composition resistor
Misc—wire and PC connector or ac twin-blade plug
  modified to fit strobe connector

31. STEREO BALANCE METER

This handy meter (Fig. 31-1) tells you electronically whether or not your stereo speakers are balanced, so there is never any doubt. It is easy to use because there is only one needle to watch.

Fig. 31-1. The Stereo Balance Meter.

With conventional output-measuring meters, either you have to compare two needles until you get identical readings on both channels or you have to switch one meter back and forth between the two channels until it reads the same for both.

Good stereo requires the careful balancing of the two speakers so that you get exactly the right amount of each signal. Merely centering the balance control on the amplifier will not guarantee that the speakers are actually delivering equal output; nor does balancing by ear ensure a perfect result because room acoustics and speaker placement can fool your ears.

The stereo balancer shown here indicates the condition of both channels simultaneously on a single meter. Instead of measuring individual channel output, it records the difference between the two channels. If the left channel is stronger than the right, the meter needle swings to the left. If the right channel is stronger, the needle moves the opposite way. You simply adjust the balance control until the needle centers, indicating that both channels are exactly equal in output. As a bonus, the meter will also check your tone controls and speaker phasing. We are indebted to Rachmiel R. Bloch who suggested the circuit on which this balancer is founded.

At the heart of the balancer is a 100-0-100 microammeter. No power supply is required since the meter is powered by the stereo signal itself. There are only three connections to make to your hi-fi set: one to the left speaker hot terminal, one to the right speaker hot terminal, and one to the common or ground terminal for both speakers. The meter can be left permanently connected since it is controlled by a push-button switch. To take a reading, you just hold the button down while adjusting the balance control. The switch keeps the meter off when it is not needed, thus saving wear on its delicate parts.

How It Works

The signals from the left and right speaker channels are fed to two miniature 100k:1k audio transformers, T1 and T2. These are wired backward to serve as step-up transformers, giving a voltage boost of 10 to 1 for greater meter sensitivity. The 1k secondaries are connected to the inputs from the speaker terminals on the amplifier and the 100k primaries are connected to transistor Q1 and the microammeter. Be sure to order these particular transformers and follow the color coding indicated in the wiring diagram of Fig. 31-2, for easy hookup of their leads. If you should use other transformers with different color coding, remember that they are wired

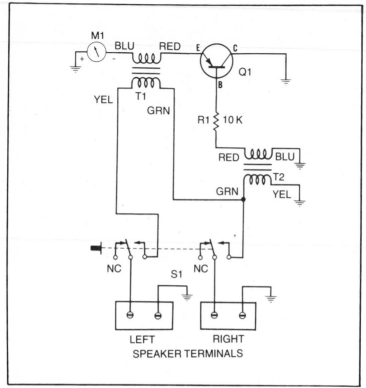

Fig. 31-2. The Stereo Balance Meter schematic.

in reverse, with the secondaries on the input side and the primaries on the output side.

In effect, the transformers are connected in series, and it is this unique arrangement that gives the balancer its simplicity. The output of transformer T1 is a difference signal representing the right channel minus the left channel. Transformer T2 provides the right channel signal only. Current polarity is determined by the relative strengths of the two channel signals. If it is positive, the meter needle swings one way; if negative, the needle swings the other way. This indicates which channel is the dominant one. When both channels are exactly equal in strength, the transistor does not conduct and no current flows through the meter, so the needle remains centered.

The transistor is an RCA germanium-type pnp 2N1303 (also Motorola HEP 629, or equivalent). Because its operation is fairly critical, it is best not to make substitutions.

Construction

The parts are housed in a small Bakelite case, measuring 1½ inches × 2½ inches × 5 inches, and having an aluminum front panel. Cut a 1½-inch × 9/16-inch rectangular opening in the center of the panel for the meter. The push-button switch is mounted in the side of the case. The transformers and other components are laid out on a strip of perforated phenolic board (Fig. 31-3) attached to the front panel with angle brackets. Use flea clips inserted in the perf board holes to make the wiring easier to do. All connections are made on top of the board except for those to the transistor leads, which extend below the board. During soldering, use pliers or a heatsink tool on the leads to avoid damage to the transistor from excessive heat.

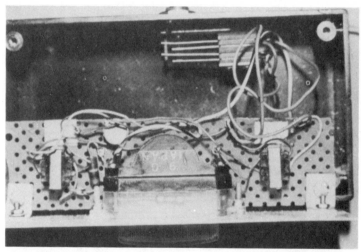

Fig. 31-3. Internal view of balance meter, showing meter and perf board.

To use the balancer, adjust the amplifier controls until the two speakers seem nearly equal in output, then press the button to take a reading. If the meter needle deflects one way or the other, touch up the balance control until it centers. The signal source should be one designed for balance tests or one in which you know both channels are approximately balanced.

In the same way, you can also check out your bass and treble tone controls to make sure that they are affecting both channels equally. If one channel has more bass or treble than the other channel, it will show up on the meter in the same way as unequal channel output. Start with the concentric knobs aligned and turn the controls up and down, taking readings at various points along the way. If the controls are tracking properly, you should get zero-deflection readings along the whole scale. If you do not, adjust the separate knobs for minimum deflection. If you cannot get the controls to track perfectly over the full range, adjust them for zero deflection at the points where you customarily keep them set. This will give you good balance at the normal listening positions.

If your stereo system has a phase switch, it, too, can be checked quickly with the meter. Simply flip the switch one way, then the other, noting the meter readings. When the speakers are out of phase, the needle will deflect; in phase, it will center. To avoid harming the meter, take all readings at moderate volume levels. A final word of caution: at extreme volume, it is possible to exceed the transistor's 30-volt breakdown rating and destroy the part.

Table 31-1 is a list of the parts required to build the Stereo Balance Meter.

Table 31-1. Stereo Balancer Parts List

```
Q1—pnp germanium transistor 2N1303 or equiv
    (Motorola HEP 629, Radio Shack 276-2007,
    or equiv)
T1, T2—1k:100k miniature audio transformers
R1—10,000-ohm, ½-W resistor
S1—dpdt push-on/push-off switch
M1—100-0-100 microammeter
Misc—case, wire, solder, etc.
```

32. SENSITIVE RF "SNIFFER"

Radio-frequency energy produced by walkie-talkie, ham, CB, and commercial transmitters—even illegal electronic "bugging" devices—is often hard to detect with a simple field-strength meter. Power levels vary widely; the needle of a meter meant to be used with a 10-watt transmitter will not even respond when exposed to the field of a 100-milliwatt transmitter—but the "Sniffer" will.

The Sniffer, shown in Fig. 32-1, is a broad-band rf detector with a built-in, adjustable-gain FET/bipolar amplifier and light-emitting diode (LED) indicator. The Sniffer can be used as a dash-top indicator for mobile radio operation. Each time you key the mike, its LED will glow, confirming that your transmitting antenna is radiating power. In the CB base station or ham shack, the handy Sniffer

Fig. 32-1. The Sensitive Rf Sniffer.

winks on when your transmitter does. You can even use it as a visual continuous-wave (cw) keying monitor because the fast-responding LED has no warm-up lag (as incandescent lamps do), so it can follow the fastest "fist." You will find many other uses for the Sniffer, too. It is ideal for checking the efficiency of shielding around power amplifiers and for locating stray sources of rf interference, such as arcing motors or high-power lamp dimmers. You can even use it to spot *static* discharges caused by charge buildup on belt-driven equipment, and with an added bit of antenna length, the Sniffer will make a good lightning detector, too.

The Circuit

The schematic diagram of the Sniffer is given in Fig. 32-2. A telescoping antenna intercepts rf energy and couples it through C1, to a detector circuit comprising an rf choke (RFC1) and germanium diode D1. (It's important to use a germanium diode for best sensitivity. Silicon diodes have a higher conduction threshold than germanium types and thus are less sensitive.) The diode is poled so that it passes only negative half-cycles to the load. Components R1 and C2 filter the rectified signal voltage to a smooth dc. The wiper of R1 picks off all or a portion of that dc voltage and applies it to the gate of field-effect transistor Q1. The FET operates at zero bias and is normally conducting, diverting bias current from the base of Q2. Hence, the LED indicator is off because Q2 is not conducting.

Assuming that the signal is of sufficient strength to bias Q1 into the pinch-off region, bias current increases in the base of Q2. As Q2 becomes forward-biased, collector current flows, exciting the LED.

The circuit gain is moderately high and constant. The sensitivity of the Sniffer is governed by the setting of R1, which determines how much of the detected signal voltage is actually applied to the gate of Q1. When the wiper of R1 is moved toward the grounded end, less of the applied signal reaches Q1. Thus, it takes a higher-level input signal to activate the circuit as the control setting is

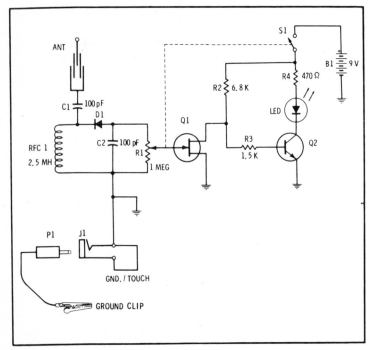

Fig. 32-2. The rf Sniffer schematic diagram.

advanced. This accounts for the wide power tolerance of the circuit and also gives you a means of estimating the output power of the transmitting source by noting the setting of R1 and seeing if a lower or higher setting produces turn-on of the LED. The control can be calibrated with a relative numeric scale, if desired, to help you make this comparison.

Upon cessation of transmission, the signal voltage at the gate of Q1 becomes inadequate to sustain pinch-off and the drain-source channel diverts current from the base of Q2, de-energizing it and the LED.

Owing to the wide frequency range in which the Sniffer may find application, the effectiveness of the antenna will often be improved through the use of a grounded-plane, such as the metal chassis of an automobile, or even the relatively large capacitance of your body to earth ground, a gnd/touch jack (J1) is provided for this purpose. It accepts a clip lead which can be snapped to a convenient ground, or the jack itself can be touched while you are using the Sniffer. Either technique provides an effective rf return, heightening the sensitivity of the circuit to lower-power fields.

Construction

The Sniffer is housed in a 3-inch × 4-inch × 1½-inch Bakelite case, and all components other than the antenna, J1, and C1 are connected together and mounted on the front panel. The arrangement of parts is not critical, but all connections should be securely made if the components are to be mounted by using the self-supporting technique shown in Fig. 32-3. Perf board may be used if you prefer.

The sensitivity control (R1) mounts directly to the front panel. Its lugs and those of its integral switch are principal tie points. The LED is fitted into a 3/16-inch hole in the panel and secured in place with a drop of cyanoacrylate adhesive. The battery mounting is similarly simplified. Since battery replacement will be infrequent (because the "on" drain is so low), the battery is secured to the panel with a drop of cyanoacrylate adhesive also. This speedy and simple mounting is quite durable, yet the bond can be readily broken when battery replacement becomes required. A drop of adhesive will then provide a "new mounting" for the fresh battery.

CAUTION: Extreme care should be exercised when using cyanoacrylate adhesive. Should any spillage occur, clean it up as soon as possible, using acetone or other suitable solvent. If you get it in your

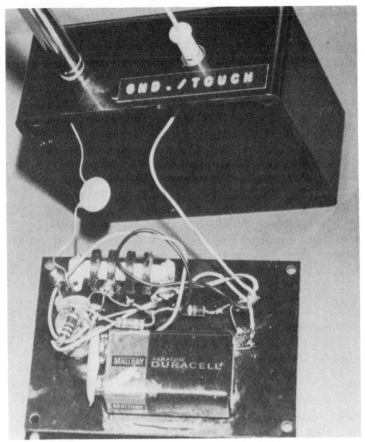

Fig. 32-3. Sniffer, with front panel removed to show component layout.

eyes, flush them immediately with water and consult your physician as soon as you can.

Use care when soldering the leads of the germanium diode (D1) and the FET (Q1), because they are quite heat-sensitive. A low-power soldering iron with a fine chisel tip is recommended. Be sure to observe correct polarity when installing the LED.

Using the Sniffer

To use the Sniffer to detect moderate power levels (between 100 milliwatts and 5 watts), extend the antenna to any length that is convenient and turn on the unit. Holding the Sniffer as far away from the suspected rf source as possible, advance the sensitivity control until the LED just winks on, and then back it off until the LED no longer burns. Bringing the Sniffer closer to the suspected source should increase the strength of the intercepted signal causing the LED to be on continuously. For detection of low-power sources (100 milliwatts or less), at frequencies above 27 megahertz, extend the antenna until its length is approximately a quarter-wavelength long.

Added sensitivity will be gained by clipping a lead wire from J1 to the transmitter, auto body, or coax shield of the transmitter's antenna line. For general-purpose use, such as the location of interference sources or the surveying of a transmitting-antenna field, extend the antenna and set R1 for maximum sensitivity. Hold the unit vertically and touch J1 while walking about. This will afford optimum sensitivity. When you are using a Sniffer for low-frequency (below 1 megahertz) detection or to detect static energy discharges (such as distant lightning), an earth ground may be patched to J1 for best sensitivity.

The frequency range of the instrument, using the components listed, lies between 5 kilohertz and 200 megahertz. Considerable variation in sensitivity occurs throughout the broad range of the instrument so that greater power is required to turn on the LED at the extremes of the frequency range. You may, if you wish, substitute a tuned LC circuit for RFC1 in order to make a highly sensitive wavemeter for a particular band. This affords a tenfold increase in sensitivity and gives quite sharp selectivity since the circuit is working into the extremely high input impedance of FET Q1,

Table 32-1. Sensitive Rf "Sniffer" Parts List

B1—9-V battery (Burgess 2N6 or equiv)
D1—germanium diode, HEP 135, Radio Shack
   276-1123, or equiv
C1, C2—100-pF disc capacitor
J1/P1—mating phone plug and jack
LED—light-emitting diode, red
Q1—n-channel field-effect transistor, HEP 801 or equiv
Q2—npn silicon transistor, Type 2N2926 or equiv
R1—miniature, 1-megohm control with spst switch
   (Mallory, Type MLC 16L-S or equiv)
R2—6800-ohm resistor   All resistors are
R3—1500-ohm resistor   ½ W ± 10%,
R4—470-ohm resistor   composition type.
S1—(part of R1)
Misc—bakelite case, antenna, battery clip, alligator
   clip, wire, solder, etc.

preserving the Q of the tuned circuit. By thus limiting the circuit bandwidth, R1 can be calibrated more precisely in terms of relative power levels.

Table 32-1 contains a list of the parts required to build the Sniffer.

33. SOUND TRANSMITTER

If you are a cassette tape buff, you have probably invested a bundle in prerecorded tapes or spent hours taping music for playback later. After all the "input" of money and time, it does not seem sensible that the output of your cassette recorder must suffer the strangulation of being reproduced by its built-in 2½-inch speaker.

With this handy little unit, shown in Fig. 33-1, you can transmit the playback audio output of your cassette recorder to any nearby AM radio, for better listening in your home, car, or camper. In addition to that, you can use the system to set up an "instant" speaker system for a house, patio, or pool party. Just place portable receivers around the area you wish to fill with sound and tune each

Fig. 33-1. The Sound Transmitter.

to the transmitting frequency of the sound transmitter. You'll have music where you want it, without the mess and bother of stringing wires, which your guests might trip over.

You can even enjoy your favorite tapes while trudging around behind a power lawnmower. Just carry a shirt-pocket AM portable radio and use an earpiece to admit the music while shutting out the racket. The sound transmitter possibilities are as limitless as your imagination.

Here's how our miniature "broadcasting station" operates: The sound signal from the cassette-unit speaker output jack is connected to the terminals on the back of the unit. The sound signal is used to modulate the amplitude of the rf carrier produced by a stable oscillator, which is operating at low power in the AM broadcast band. This is exactly the same form of modulation that is used by commercial broadcasting stations. To ensure drift- and distortion-free operation of the transmitter, the oscillator frequency is con-

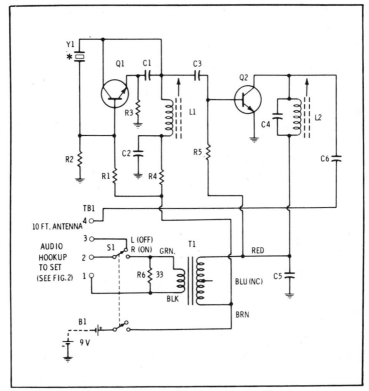

Fig. 33-2. The schematic diagram of the Sound Transmitter.

trolled by a precision quartz crystal. The crystal frequency should be selected so that your unit operates on a clear frequency, near the upper end of the broadcast band, where there is no reception in your area. We chose 1500 kilohertz as the frequency of operation, since it is a relatively clear channel; 1510 and 1520 kilohertz are good alternates. Depending upon the AM band assignments in your area, pick one of these frequencies and order the crystal from the source specified in the parts list.

Power switch S1 has a dual function; it controls power application and transfers the audio signal from the speaker of a unit not equipped with an output jack into the modulated rf amplifier stage of the sound transmitter (Fig. 33-2). The audio applied to terminals 1 and 2 of the unit is coupled to the input transformer when S1 is in the R (remote) position. A 33-ohm resistor across the transformer input ensures a constant input impedance at most audio levels. The output of the cassette player can alternatively be connected to terminal 3 of the unit, so that its built-in speaker is restored when the switch is set to the L (local) position. This is true only where your player has no output jack, and will necessitate some minor surgery at the speaker terminals.

The Circuit

At 1500 kilohertz, crystal oscillator Q1 produces a stable rf signal which is coupled to the base of rf amplifier Q2. The amplified output of transistor Q2 appears across resonant circuit L2 and C4 and is coupled to the antenna lead through C6. It is important that the overall antenna length not exceed 10 feet. A longer antenna violates Federal Communications Commission (FCC) regulations and will also result in poor operating efficiency of your unit.

The incoming audio is applied to the dc collector voltage supplied to Q2, and modulation percentages up to 100 percent are possible, without distortion. The stability of the crystal oscillator makes for clear, crisp audio reproduction through any nearby AM receiver and far surpasses the performance of any self-controlled oscillator. This makes the slight extra expense of the crystal worthwhile.

The amplitude-modulated rf output of Q2 is radiated by the antenna lead, providing a strong AM signal within 30 feet of the antenna. The level of modulation is set by simply adjusting the volume control on the tape player. Thereafter, audio adjustments can be made by using the volume control on the AM receiver at the remote location.

170

Construction

The circuit is contained in a small metal box (Fig. 33-3). All components are mounted on a perforated phenolic board and soldered to small clips. The photographs clearly show the parts layout. When you have assembled the unit, apply power and listen for its signal on a nearby AM receiver. Using a nonmetallic hex alignment tool, adjust the ferrite core of coil L2 for maximum output, as heard on the nearby receiver.

Fig. 33-3. The Sound Transmitter, with cover removed to show component layout.

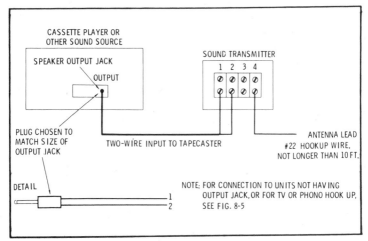

Fig. 33-4. Hook-up to sources with an audio output jack.

The output power of our transmitter is sufficient to ensure excellent reception within thirty feet from the set, but at the same time it is also low enough to be below the maximum rf field intensity specified by the FCC for such low-power devices. (See Part 15 of the FCC Regulations.)

This sound transmitter can also be used with your hi-fi equipment and portable phonograph or for other similar wireless applications. As long as the source provides an audio output, the sound transmitter will broadcast it for your listening enjoyment. Figures 33-4 and 33-5 show the two alternative methods of connecting the sound transmitter to your sound source.

A list of parts required to build the Sound Transmitter is given in Table 33-1.

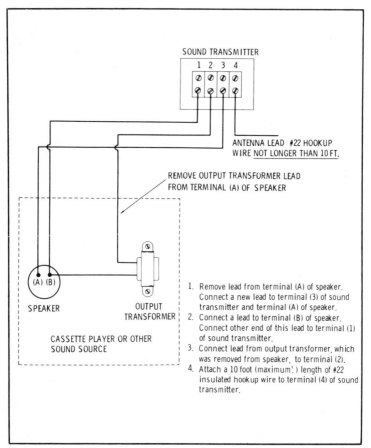

Fig. 33-5. Hook-up to sources without an audio output jack.

Table 33-1. Sound Transmitter Parts List

B1—9-V battery (Burgess 2N6, or equiv)
C1, C3, C6—100-pF disc capacitor
C2, C5—.01-μF disc capacitor
C4—300-pF disc capacitor
Q1, Q2—npn silicon transistor, Type 2N3663, or equiv
   (Radio Shack 276-2010)
R1—470k, ½-W resistor
R2—4.7-megohm, ½-W resistor
R3—330-ohm, ½-W resistor   All resistors are
R4—1000-ohm, ½-W resistor   ± 10%
R5—22k, ½-W resistor
R6—33-ohm, ½-W resistor
S1—dpdt toggle switch
T1—transistor transformer (Argonne, Type AR-164, or equiv)
L1—160-μH adjustable coil (J. W. Miller, Type 23A224RPC,
   or equiv)
L2—43.5-μH adjustable coil (J. W. Miller, Type 23A685RPC,
   or equiv)
Y1—quartz crystal, Type TC-23, 1500, 1510, or 1520 kHz;
   select *one* frequency depending upon AM broadcast band
   assignments in your area. Specify frequency and order
   Type TC-23 from: Texas Crystals, 1000 Crystal Drive, Fort Myers,
   FL 33901
Misc—aluminum case (Premier PMC-1006, Radio Shack
   270-238, or equiv) 5¼″ × 3″ × 2⅛″; ½″ spacers (4)

34. WIRED RADIO REMOTE SPEAKER

When your house was wired for ac power, you got an unexpected bonus. Those same wires snaking through your walls can also be used to carry music from your hi-fi set to any room where you want an extension speaker. The system is called multiplexing—making one pair of wires do two jobs at once. With this arrangement, you don't have to string separate wires to permanently installed remote speakers. You can carry a single, lightweight speaker from one location to another as you need it—in a bedroom or playroom, on the back terrace—anywhere you have an electrical outlet.

The system consists of two easily wired electronic units—one to transmit the audio signal from your hi-fi set to the ac line (Fig. 34-1) and one to receive the signal at the remote speaker (Fig. 34-2). The transmitter connects to the speaker terminals on your amplifier and plugs into a nearby wall outlet. The receiver plugs into a distant outlet and two wires go to the speaker. These are the only connections you make.

You can build as many receiver units as you like. They will all

Fig. 34-1. The transmitter.

Fig. 34-2. The receiver.

pick up signals from the same transmitter so you can have speakers playing in various parts of the house simultaneously if you wish. Each receiver has its own volume control so you can adjust it independently of other units.

The Circuit

Both the transmitter and receiver incorporate identical 455-kHz i-f transformers. These are detuned through capacitors to a frequency of about 400 kHz. This is the carrier frequency that rides the ac line to transmit the audio signals over the wires. Because it is outside the range of commercial AM broadcasts (550 to 1600 kHz), there is no interference with radio signals. The transmitter and receiver are connected to the ac through coupling capacitors that isolated their circuits from the line voltage while allowing the carrier signals to pass through.

Construction

To simplify construction, the receiver (Fig. 34-3) uses a pre-wired 3-watt audio amplifier module available from Lafayette Radio, 111 Jericho Turnpike, Syosset, NY 11791. This is a printed-circuit board requiring only seven connections—two on the input side, using shielded cable; two on the output side to the speaker terminal

Fig. 34-3. The receiver with the cover removed.

block; two to ground; and one to the power supply. Follow the instructions packed with the amplifier for making these connections. The wiring diagram accompanying the amplifier shows the circuit board as it actually appears and will also help to guide you in locating the proper hookup points.

The receiver unit (Fig. 34-3) is housed in a 7-inch × 5-inch × 3-inch gray metal case. The smaller transmitter, shown in Fig. 34-4, requires only a 5¼-inch × 3-inch × 2-inch box. In both instances, the power-supply components are installed directly on the inside of the metal boxes. Small parts are wired on pieces of phenolic perforated board with flea clips used for making the connections. Mount these boards on standoff spacers so that they will not touch the sides of the boxes. The i-f transformers are secured to the perforated board with contact cement. Use spacers when mounting the amplifier module in the receiver and be careful that the foil surface of the printed circuit board does not touch other parts.

Fig. 34-4. The transmitter with the cover removed.

Using the Remote Speaker

The only controls involved are a simple on-off switch for the transmitter and a combination on-off switch and volume control for the receiver. After the parts are assembled, it may be necessary to adjust the i-f transformers for good alignment. Plug in both units and turn them on. Set your radio or hi-fi set to a comfortable volume level; then switch off or disconnect your main speakers to be sure you are listening only to sound from the remote speaker. If you

176

cannot hear the transmitter signal, tune IFT2, the i-f transformer in the receiver, by turning the core in or out until you hear the signal.

If this still does not bring in the sound, the transmitter frequency may be slightly outside the tuning range of the receiver. In this case, tune the transmitter i-f transformer, IFT1, by turning the core in or out until the receiver picks up the signal. You may have to do this several times, adjusting first one and then the other, until the transmitter and receiver are perfectly aligned.

The sound should be clean and free from hum. If any hum is noticeable, turn down the volume on the receiver and turn up the volume on the hi-fi set instead. This should eliminate any problems. Since the system is monaural, turn the mode selector on your hi-fi set to the monaural position when you are listening to the remote speaker so that you'll get a mixture of both channels.

The schematic drawings are shown in Figs. 34-5 and 34-6. The parts lists for the transmitter and receiver are in Tables 34-1 and 34-2, respectively.

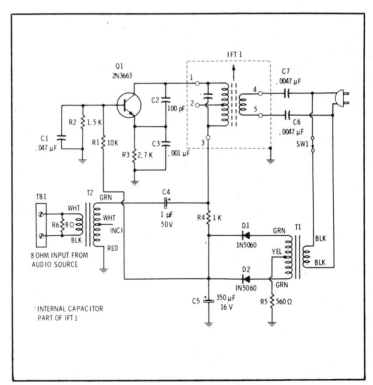

Fig. 34-5. Schematic diagram of the transmitter.

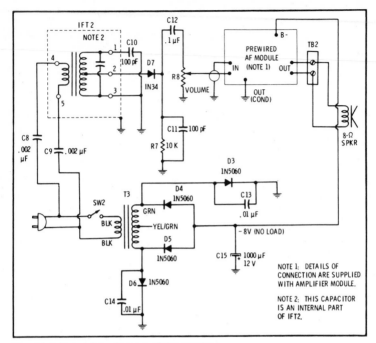

Fig. 34-6. Schematic diagram of the receiver.

Table 34-1. Transmitter Parts List

C1—.047-μF disc capacitor
C2—100-pF disc capacitor
C3—.001-μF, disc capacitor
C4—1-μF, 50-Vdc electrolytic capacitor
C5—350-μF, 16-Vdc electrolytic capacitor
C6, C7—.0047-μF, 1000-Vdc disc capacitors
D1, D2—GE silicon diodes Type 1N5060 (Motorola HEP 156, Radio Shack 276-1104, or equiv)
Q1—GE transistor, Type 2N3663 (Motorola HEP 50, Radio Shack 276-2010, or equiv)
R1—10k, ½-W ± 10% resistor
R2—1.5k, ½-W ± 10% resistor
R3—2.7k, ½-W ± 10% resistor
R4—1k, ½-W ± 10% resistor
R5—560-ohm, ½-W ± 10% resistor
R6—8.1-ohm, ½-W ± 10% resistor
SW1—spst toggle switch
T1—miniature control transformer with 117-V primary, 24-Vdc secondary at 300 mA (Radio Shack 273-1386)
T2—miniature audio output transformer with 1000-ohm primary and 8-ohm secondary (Radio Shack 273-1380)
IFT1—455-kHz i-f transformer
TB1—2-terminal barrier strip (Radio Shack 274-656)
Case—5¼″ × 3″ × 2⅛″ gray metal box (Premier PMC-1006, Radio Shack 270-238, or equiv)
Misc—phenolic perforated board for chassis, flea clips, 4 spacers, terminal strip, 6-32 hardware, hookup wire

Table 34-2. Receiver Parts List

C8, C9—.002-μF, 1000-Vdc ceramic disc capacitors
C10, C11—100-pF, 1000-Vdc disc capacitors
C12—.1-μF disc capacitor (25 V minimum)
C13, C14—.01-μF disc capacitors
C15—1000-μF, 12-Vdc electrolytic capacitor
D3, D4, D5, D6—GE silicon diodes, Type 1N5060
   (Motorola HEP 156, Radio Shack 276-1104,
   or equiv)
D7—GE signal diode, Type 1N34 (Motorola HEP 134, Radio
   Shack 276-1123, or equiv)
IFT2—455-kHz i-f transformer
R7—1-megohm, ½-W resistor
R8—10k volume control, linear or audio taper, with spst
   switch on back
SW2—spst switch attached to R8
T3—filament transformer with 117-Vac primary, 6.3-Vac
   secondary, 3-amperes, center-tapped
TB2—2-terminal barrier strip (Radio Shack 273-1380)
Amplifier section—prewired 3-W transistor audio-amplifier
   module
Case—7″ × 5″ × 3″ gray metal box (Premier Type
   PMC-1008 or equiv)
Misc—shielded lead, phenolic perforated board for
   chassis, flea clips, 4 spacers, terminal strip, 6-32 hardware,
   hookup wire, line cord, rubber grommet

35. METAL DETECTOR

If you've ever tried to find a buried pipe or electrical conduit, or wondered which way a BX cable travelled through a wall, you have encountered just two of the many practical uses for a metal detector. There are fun applications, too. Beachcombing and treasure hunting with a metal detector (Fig. 35-1) are just some of the leisure time activities that will enhance your vacation enjoyment. Countless games can be devised to keep youngsters happily occupied; for example, bury a few coins around your property and turn the kids loose to find them.

Whether you build this metal detector for fun or practical use, you will find it simple to assemble and use. Part of this simplicity comes about through the use of an AM transistor portable radio as one of the key building blocks. The rest of the circuit is simply a one-transistor oscillator that is tunable across the 550- to 1600-kilohertz AM broadcast band. You can assemble this metal detector (Fig. 35-2) in a single evening.

Fig. 35-1. Debbie Graf using the Metal Detector.

How the Metal Detector Works

It is a well-known fact that if two rf oscillator signals are mixed and applied to a common detector, *sum* and *difference* frequencies are generated as the two signals beat together. (This is what makes the superheterodyne receiver possible.) If the two oscillator signals are tuned to exactly the same frequency, there is no difference frequency. This is called *zero-beat*.

When one of the oscillators (the search oscillator) is arranged so that its tuned-circuit coil can be brought near a metal object, the metal will slightly alter the coil inductance, thus shifting that oscillator's signal frequency. As the signal moves out of zero-beat with the other (reference) oscillator, an audible note will be heard. This is the principle behind the metal detector.

To keep costs down, this metal detector has only a search oscillator. The "reference oscillator" is a fairly weak AM radio station signal, tuned-in on a transistor portable radio. By zero-beating the search oscillator's signal against the radio station's

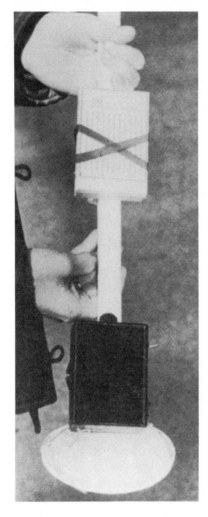

Fig. 35-2. The Metal Detector.

signal and then positioning the search coil over a buried metal object, an audible note will be heard through the radio's speaker.

The Circuit

The battery-powered search oscillator shown in Fig. 35-3 is a Colpitts type, in which feedback to sustain oscillation is applied from collector to emitter through capacitive divider C3, C4. The base of Q1 is at rf ground, through the bypassing action of C2. The return end of the tank coil is bypassed by C5. Operating bias is provided by base resistor R1 and emitter resistor R2.

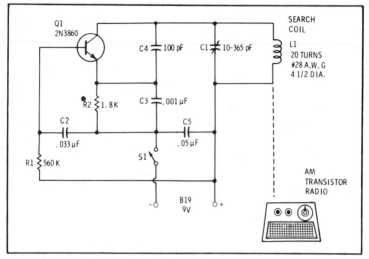

Fig. 35-3. The Metal Detector schematic diagram.

This oscillator allows the use of a two-terminal coil, simplifying the construction of the search coil. The tuning capacitor, C1, is the type ordinarily found in transistor portables, although an air variable or mica type can be substituted.

Construction

The search oscillator circuitry is assembled within a small plastic box (Fig. 35-4) (approximately 2¾ inches × 3¾ inches × 1¼ inches), which also contains the battery and power switch. Tuning capacitor C1 is fastened to a side of the box for convenience in setting the frequency. Wiring is point to point and is not especially critical, although connections should be securely made and components placed so that vibration is minimized.

The search coil is wound with cloth-covered or plain-enamel No. 28 AWG copper wire. For simplicity, we used the body of a 4¼-inch-diameter plastic refrigerator dish as the form, winding 20 turns around the lip end. Then the coil was slid off the dish and taped together at several points, forming a 4¼-inch-diameter "doughnut." The coil was then pressed into the sealing ridge of the refrigerator dish cover and secured in place with adhesive. The result is a rugged assembly that is easily mounted, protects the coil, and retains its shape.

Use a 5-foot length of broomstick or a 1-inch dowel as the "handle" for mounting the metal detector components. Cut one end

182

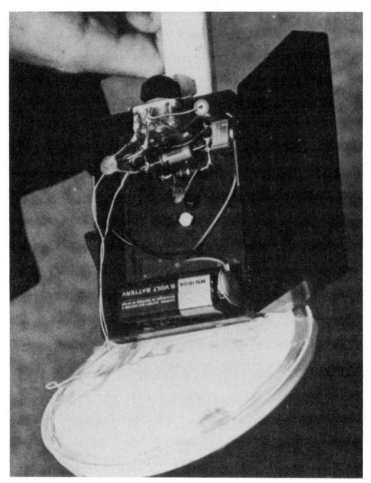

Fig. 35-4. The search oscillator.

of the handle at a 45° angle to accept the search coil. This will allow you to hold the handle at a comfortable angle yet keep the coil in a horizontal plane parallel to the ground surface.

Secure the search oscillator box to the handle with wood screws. Place it not more than 1 inch above the angled end of the handle and in line with the angular portion that extends farthest outward.

Next, mount the search coil assembly. Use a brass screw and a plastic or fiber washer so that no ferrous metal will be within the coil field. Connect the two wires of the coil to the tuning capacitor terminals in order to complete the search oscillator circuit.

Snap a 9-volt battery into place in the search oscillator and position your AM band transistor portable radio nearby. Turn both on and tune the radio to a moderately weak station in the 550- to 800-kilohertz range. Slowly tune capacitor C1 across its range until you hear a whistling tone in the speaker, signifying that the oscillator is working and beating with the radio station's signal. (If you are unsuccessful, carefully recheck your wiring connections and soldering to find the error or fault that is preventing oscillation.)

After this quick check, mount the transistor portable on the stick, if you wish, about 6 to 12 inches above the search oscillator; the method you use to mount it depends on the construction of your radio, but it can be as simple as the two elastic bands we used.

Using the Metal Detector

With the radio and the search oscillator on, hold the stick vertically and keep the search coil away from any metal objects. Now, tune the radio to a moderately weak station and adjust C1 until you hear a beat note. Fine-tune C1 to zero beat the search oscillator at the radio station frequency.

Sweep the search coil past a metal object, and you will hear a note of rising pitch that falls back to zero beat after the coil has passed the object. Now, try sweeping the search coil along the ground to locate a buried object. You should be able to detect a steel key or heavy bolt lying on the ground surface, but beneath the grass. Buried objects of large size are readily detectable at depths up to one foot in very dry soil. Sensitivity falls off sharply in rain-saturated soil. Practice and experimentation will gradually improve your ability not only to detect objects but to judge their size and depth by the characteristics of the pitch change.

Table 35-1. Metal Detector Parts List

B1—9-volt battery (Burgess 2N6 or equiv)
C1—miniature variable capacitor, 365-pF max
C2—.033-μF Mylar capacitor
C3—.001-μF Mylar capacitor
C4—100-pF disc capacitor
C5—.05-μF Mylar capacitor
L1—20 turns No. 28 A.W.G. cloth-covered or plain enameled wire on 4½" diameter form
Q1—npn transistor, Type 2N3860 (Radio Shack 276-2010)
R1—560,000-ohm, ½-W ± 10% resistor
R2—1800-ohm, ½-W ± 10% resistor
S1—spst switch

Lots of home and hobby activities call for accurate time-keeping: timing games; boiling an egg; cooking a roast; giving yeast time to work; curing paint and adhesives; exposing, developing, and washing prints in a photolab; timing "on-air" transmissions on CB and ham transmitters so they are within the legal limit, setting a "gab" limit on phone calls; timing a speed run along a measured mile—all these situations and hundreds more that you can think of, call for measuring intervals of varying lengths with good accuracy. Few "clockwork" timers can handle periods of a few seconds *and* periods of two or more hours. Those that can are usually ac operated and costly.

The General-Purpose Timer shown in Fig. 36-1 should prove versatile enough to meet all your interval timing needs. It is com-

Fig. 36-1. The General-Purpose Timer.

pact, lightweight, battery-powered, and low in cost. Set it to time a few seconds, several minutes, or a couple of hours. Press the start button and the integrated-circuit "works" begin measuring the interval you want. When the last second elapses, the timer sounds a 10-second "beep," then automatically shuts down to conserve battery power. Want to change intervals after one is started? Press the zero button, dial in the new period, and push the start button. It is as simple as that! Battery drain is below 4 milliamperes throughout a timing cycle, so the cost of the operation is negligible.

The Circuit

A Type-555 integrated-circuit timer is the "movement" of our timer. It is a monolithic, highly stable controller capable of producing very precise time delays with a bare minimum of external parts. Best of all, it works with common aluminum electrolytic capacitors as timing elements, rather than the costly, but less leaky, tantalum types. This makes it possible to use multiple timing ranges to cover the periods you have to measure most often. Figure 36-2 shows the circuit of the General-Purpose Timer. The 555 (IC1) here functions as a monostable, or "one-shot," multivibrator. Pin 2 is biased positively through R1, providing a reference voltage to an internal comparator and holding an internal flip-flop "off." The range switch, S1, selects and connects a capacitor between pin 6 and ground. The time set control, J2, and resistor R3 supply current to that point, proportional to the setting of R2. But, in the "off" state, an internal transistor of the 555 chip shorts the timing capacitor, so nothing happens.

However, when start switch S3 is momentarily pressed, pin 2 is pulsed negatively. This resets the internal flip-flop, which releases the short circuit across the external capacitor and drives the 555 output (pin 3) to +9 volts. The selected timing capacitor now commences to charge through R2, R3. When the voltage across the capacitor reaches approximately 6 volts, the 555 comparator resets the flip-flop, driving the output (pin 3) to zero and also discharging the timing capacitor. Thus, operating as a true "one-shot," the circuit was triggered by a negative-going input; its normally zero output remained high throughout the timing cycle and then returned to the stable (zero) state upon expiration of the required period.

Once triggered, the circuit will remain in the timing state until the set time is elapsed. To cancel an interval and reset the time, zero push button S4 is momentarily depressed. This instantly

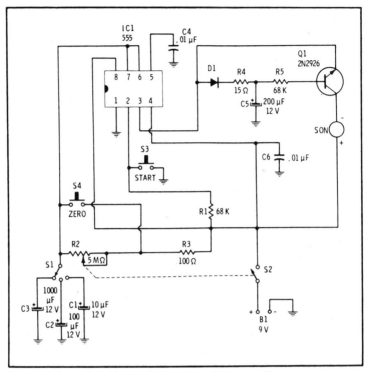

Fig. 36-2. Schematic diagram for the General-Purpose Timer.

"times out" the circuit, so you can change the setting of R2 and begin a new cycle by pressing S3.

The timer uses a novel sound-signal circuit consisting of a Sonalert® controlled by transistor switch, Q1. When the timer is first turned on, the base of Q1 is at zero potential and the emitter is grounded through the output, pin 3, of the 555. Thus, Q1 does not conduct and the Sonalert is silent.

During the timing cycle, the output pin of the 555 rises to +9 volts. This is applied through diode D1 and limiting resistor R4 to capacitor C5. This capacitor stores a considerable charge which is applied to the base of Q1 through R5. But, Q1 still cannot conduct because its emitter is held at the +9-volt level of output pin 3 of the 555 throughout the timing cycle.

However, upon expiration of the delay, when the 555 output falls to ground, the emitter of Q1 is also grounded. Instantly, the charge stored in C5 biases Q1 on, supplying a path to the Sonalert which is completed to ground through the 555 output.

Transistor Q1 continues to conduct until the charge on C5 has bled off. (This takes approximately 10 to 12 seconds with the values shown for C5 and R5.) When Q1 ceases to conduct, the Sonalert is silenced and the timer waits for you to start another cycle.

If you wish, the circuit can be further elaborated to include switching components, such as a relay or triac, so that power can be controlled to an external load. As this will depend largely upon your specific application, the exact design is up to you. You can, however, fit a small power-line-operated dc supply into the unit and parallel a sensitive reed relay coil across the Sonalert to control a husky power relay or triac. Limitless possibilities are open to you with this intriguing circuit!

Construction

Our battery-powered General-Purpose Timer (Fig. 36-3) is packaged in a 7-inch × 5-inch × 3-inch miniature aluminum case. A smaller enclosure can be used, but the larger case gives exterior space for an easy-to-read calibration scale.

The Sonalert, control R2; switches S2, S3, and S4; and battery B1 are mounted directly on the case as shown in the accompanying photos of Fig. 36-4. All other components fit on a handy, prefab printed-circuit board. The board is meant to simplify working with 14-lead DIP-type ICs. Since the 555 is an 8-lead DIP-type, there is extra space and unused pads on the board to serve as tie points for other components. The IC simply drops into place on the predrilled board and is soldered on the foil side, using a low-heat iron. The wiring of the other parts is shown in the photos.

Your choice of timing ranges is not limited to three; you can add more capacitors to create as many selectable ranges as you wish. Or, you can vary the sizes of the capacitors to obtain the intervals that best suit your needs. However, if the value of any of the timing capacitors (C1, C2, or C3) or the timing resistor (R2) is changed after the initial calibration has been completed, it will be necessary to recalibrate the timer. Also, as time goes by, the value of these components may change, necessitating recalibration.

Calibration

You can make a calibrated scale for your timer by using a clock or stopwatch to check seconds or minutes. But, calibrating a scale (Fig. 36-5), in divisions up to 2 hours is tedious by this method.

For accuracy in high time-scale calibration, we suggest a different approach. First, use a compass to inscribe a circle on a scrap

Fig. 36-3. Internal view showing switch and component board layout.

sheet of paper. Next, divide the circle into 5-degree segments with a protractor. Mark the five-degree points along the circle, from 0 to 360. Cut out the center of the paper and fit the scale over the shaft of time set control R2. Rotate the shaft to the point just past the closure of the power switch, S2. Affix a pointer knob to the shaft of R2 and rotate the scale until the 0° mark is aligned with the pointer. Temporarily tape the scale to the case. Now, turn the time set control until its knob lines up with the 30° mark. Time the interval with a clock or stopwatch and note the elapsed time. Move to the 60° mark, time the interval, and note it. Next, move to the 90° mark

Fig. 36-4. Close-up view of the PC board.

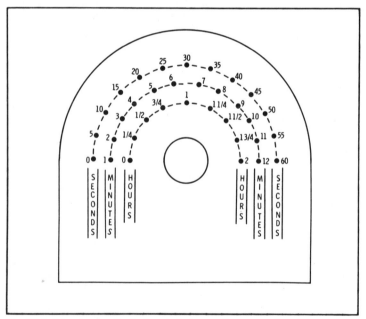

Fig. 36-5. Sample of calibrated scale to be used as example only.

and repeat. Do this for each 30° increment, up to the limit of control rotation (approximately 270°, allowing for the control rotation needed to actuate the switch). You now have nine time intervals, corresponding to exact rotational degree settings of R2.

Obtain a sheet of semilog graph paper (Dietzgen, No. 340R-L210) from a local art or drafting supply store. This printed graph is semilogarithmic along the lengthwise (11-inch) dimension but linear (divided into equal increments) along the widthwise (8½-inch) dimension.

Next, on the widthwise (8½-inch) linear scale of the paper, mark the degrees in ascending number from 0, in increments of 20 degrees, up to 280°. Now, mark off the hours scale on the semilog (lengthwise) axis of the paper. Since ¼ hour (15 minutes) is the lowest usable point on the hours range, mark the fourth major division from the left edge as "¼." Count off four more major divisions to the right and mark "½." Count another four and mark "¾"; another four, and "1." Now, count two divisions and mark "2."

Table 36-1. General-Purpose Timer Parts List

B1—9-volt battery (Burgess 2N6 or equiv)
C1—10-μF, 12-Vdc electrolytic capacitor
C2—100-μF, 12-Vdc electrolytic capacitor
C3—1000-μF, 12-Vdc electrolytic capacitor
C4, C6—.01-μF disc or Mylar capacitor
C5—200-μF, 12-Vdc electrolytic capacitor
D1—general-purpose germanium diode, Type 1N60 (Radio Shack 276-1123)
IC1—Type 555 IC timer (dual in-line package, DIP) (Radio Shack 276-1723)
Q1—npn transistor, Type 2N2926
R1, R5—68,000-ohm resistors
R2—5-megohm control, with spst switch   } All resistors are
R3—100-ohm resistor   ½-W ± 10%,
R4—15-ohm resistor
SON—Sonalert (P. R. Mallory Co., Type SC-628 or Radio Shack 273-060)
S1—2-pole, 6-position rotary switch (only one pole is used, but any number of positions can be used, depending on required ranges
S2—(part of R2)
S3, S4—spst normally open miniature push-button switch Radio Shack 275-1547)
Misc—miniature aluminum case, 7″ × 5″ × 3″ (Premier Metal Co., type PMC-1003 or equiv); dual in-line IC socket adapter; pointer knobs (2); battery clip; wire; solder; etc.

Count another two divisions and mark "3," etc. When you're finished, you'll have a graph of linear degrees versus an exponential scale of time in hours.

Now, find each of the nine points you actually timed and mark them on the paper at the intersection of degrees versus time. Then, connect the nine points with a smooth curve. You now have a graphic plot of time versus degrees for any incremental rotation of R2 in the hours scale. Simply transfer the time from the graph for each 5° marking on the paper scale affixed to the unit. This will give you an accurate hours scale.

Should you wish to add time ranges by using additional capacitors on range switch S1, you need simply lay out another time scale on the hours dimension of the graph and use the same method to obtain an accurate plot of time versus degrees for the new range capacitor.

37. "SMART" SOUND-ACTIVATED SWITCH

Most sound-activated switches are suitable for library-quiet areas only, because they react to noise like a nervous cat. That's because they respond only to the *amplitude* of an applied signal. However, this "Smart" Sound-Activated Switch (Fig. 37-1) has the

Fig. 37-1. The "Smart" Sound-Activated Switch.

ability to distinguish between most ordinary background noises, such as speech or music, and the desired actuating audio signal. What's more, it has a "memory." It switches from "off" to "on" and stays put until another signal tells it to switch from "on" to "off." You can use it to squelch a tv set or radio during annoying commercials or use it for "no-hands" control of toys, games, and photographic lights. The built-in relay can switch currents of up to 1 ampere at 125-volts ac and can control a husky external relay or triac for switching larger currents.

The Circuit

Figure 37-2 is a schematic of the "Smart" Sound-Activated Switch. A prefabricated audio amplifier module raises the level of audio voltages produced by a 100-ohm loudspeaker, SP1, used here as an efficient dynamic microphone. You can use an 8-ohm speaker here if you use an impedance-matching transformer. An output transformer can be used by connecting the 8-ohm secondary to the speaker and the 1000-ohm primary to the amplifier input. The amplifier level is adjusted by gain control R1.

The transformed output of the amplifier module is coupled to the base of pnp transistor Q1 from the low-impedance (8-ohm) winding of the built-in output transformer. (In the module used by the authors, one end of the output transformer secondary winding is directly connected to +9 volts through the module printed-circuit board. If another module is substituted, check its schematic and make its secondary winding connections agree with the schematic given in Fig. 37-2.

Normally, Q1 is reverse-biased and nonconducting. However, on negative excursions of the applied audio, Q1 switches on, applying a positive current pulse that charges C1, through D1 and R4. Although these components are in parallel with collector load resistor R3, diode D1 prevents C1 from discharging through R3 and R4 when Q1 switches off. Instead, discharge resistor R5 slowly bleeds off the charge accumulated in C1.

It takes about 1.5 seconds to charge C1 through R4, and about 5 seconds to fully discharge C1. This timing ratio helps to immunize the circuit against unwanted sounds. The emitter of unijunction transistor Q2 is connected across C1 and R5. Normally, this path is nonconducting. However, if a signal is both *loud* enough and *long* enough to charge C1 faster than R5 can discharge it, the voltage reaches the breakover point of unijunction transistor Q2 and it conducts. In doing this, Q2 allows C1 to discharge and develop a

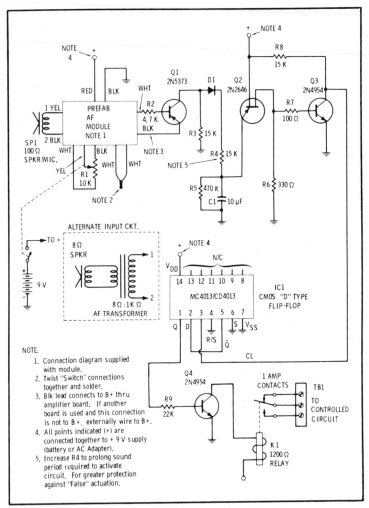

Fig. 37-2. Schematic for the "Smart" Sound-Activated Switch.

single, brief-duration pulse across the base 1 resistor, R1. This pulse instantly drives Q3 (which is normally off) into conduction, then back to the off state. The result is a negative-going pulse, returning very rapidly to +9 volts (that normally appears at the collector of Q3), and is applied to the clock input (pin 3) of a "D-type" CMOS flip-flop, IC1.

CMOS logic was chosen because it provides exceptional power gain with very high noise immunity. This means low current drain and little chance that the IC will be falsely triggered by transients.

In the normal state, the clock input of flip-flop IC1 senses the steady +9 volts at the collector of Q3. This does not trigger the flip-flop. However, when Q3 rapidly turns on and off in response to a pulse from Q2, the clock input "sees" a rapid pull-down to ground, followed by a steep rise to +9 volts as Q3 turns off. It is this sharply rising wavefront that triggers the flip-flop, causing it to switch to its other stable state.

The IC is arranged as a "toggle" by connecting the data input (pin 2) to the $\overline{Q}$ output (pin 5) and grounding both the set (pin 6) and reset (pin 4) inputs. The Q output (pin 1) advances from "off" to "on" on the first pulse, and from "on" to "off" on the next pulse. Both states are stable while power is applied.

The Q output of IC1 is supplied to the base of relay driver Q4, through resistor R9. When the Q output is high, Q4 is forward-biased and pulls in relay K1. When the Q output is low, Q4 is reverse-biased and relay K1 drops out. Terminals on the back of the unit allow external circuits to be made and broken through the contacts of K1.

The discriminatory power of the circuit relies, in part, on the fact that speech or musical sounds possess lesser energy than a sustained note or shout. In areas subject to loud, long noises, the unit may be falsely actuated. In most home environments, the unit will give more than satisfactory service. Greater discrimination against unwanted noise can be achieved by increasing the value of R4.

Construction

The "smart" sound-activated switch is housed in a Radio Shack Perfbox®. This case has predrilled openings to fit a 3-inch-diameter speaker and has a removable perforated board back. The 100-ohm speaker (SP1) is simply glued in place behind the speaker openings. We sparingly applied a cyanoacrylate adhesive to the speaker rim. Don't get any on the paper core or you will ruin the speaker!

The gain control, R1 (containing switch S1), is mounted in a hole drilled below the speaker (Fig. 37-3) and reamed to size. Other components are mounted on the back of the perf board and on an IC adapter screwed to the case end. The amplifier module is screwed to the board, and terminal strips are secured under the screws to mount other components. Wiring is point to point and can be seen from the accompanying photo. The schematic, shown in Fig. 37-2, gives the colors of the wires that come attached to the amplifier module. If yours differ, consult the schematic supplied by the mod-

Fig. 37-3. The switch with the rear cover removed showing PC boards.

ule manufacturer and wire accordingly. Be sure to connect one end
of the output-transformer secondary winding to B+, unless the
module comes already wired that way. Check carefully.

Your choice of a power supply depends on the kind of service your unit must give. For occasional use, you can use a husky battery supply made up of six "AA" 1.5-volt cells in series, installed in a holder secured to the perf board back of the unit. Alternatively, you can use a universal ac adapter (Fig. 37-4). This compact, husky supply comes prepackaged in its own case which plugs into the ac power outlet. The line connection is safely made and the supply is enclosed, so there is no shock hazard. You can fit a jack to the unit, matching one of the types furnished with the adapter, so that you can choose battery or ac operation, with ease.

(CAUTION: Don't reverse the polarity of the adapter dc output or you may damage the amplifier module of your sound switch.)

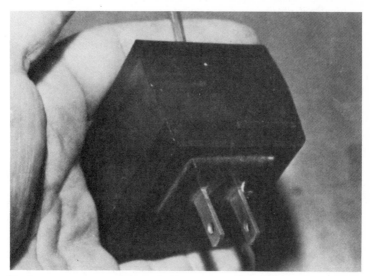

Fig. 37-4. A universal dc power supply.

Using the "Smart" Sound-Activated Switch

Turn the unit on and set its control to a point where it is insensitive to slight background "noise" (low speech or music at a fairly low level, at least 10 feet from the unit). Now, whistle a sustained note or say "ah-h-h," holding it for at least two seconds. The switch should operate. If it doesn't, try a little higher setting of the gain control.

With a little practice in setting it up, you will be able to activate your unit from fairly long distances. If you find that high-level noise is tripping the unit, lengthen the turn-on time by increasing the value of R4. This will mean holding the activating sound for a bit longer, but it will allow use of the unit in noisier environments.

Table 37-1. "Smart" Sound-Activated Switch Parts List

C1—10-μF, 25-Vdc electrolytic capacitor (Radio Shack 272-1013)

D1—general-purpose germanium diode, 1N60 or equiv (Radio Shack 276-1123)

IC1—CMOS "D-type" flip-flop integrated circuit (RCA CD4013, Motorola MC 4013, or equiv)

K1—miniature spdt relay (Radio Shack 275-004, 275-003 or equiv)

R1—10,000-ohm, ½-W potentiometer with spst switch (S1) (Radio Shack 271-1443 or equiv)

R2—4700-ohm resistor

R3, R4, R8—15,000-ohm resistors

R5—470,000-ohm resistor   All resistors are ½-W

  R6—330-ohm resistor   ± 10%, composition type.

R7—100-ohm resistor

R9—22,000-ohm resistor

Power—Six "AA" cells in series to form a 9-volt supply. Mount in holder on perforated board back of unit. on sound-switch unit so adapter can be disconnected.

Alternative ac Supply: universal ac adapter. Use mating jack or connector

Misc—Perfbox®, 3¾″ × 6⅛″ × 2″; terminal strips; screw terminal strip; dual in-line IC socket board adapter

SPKR—100-ohm impedance, 3″ dia. speaker. (Calectro S2-206 or equiv)

Q1—pnp silicon transistor, Type 2N5323 or equiv

Q2—unijunction transistor 2N2646 or equiv

Q3, Q4—npn silicon transistor 2N4954 or equiv (Radio Shack 276-2009)

AMPL—5-transistor amplifier module, 260-mW

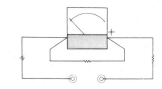

Safety Electronics

Safety, according to *Webster's* Dictionary, means being free from danger, injury, or damage. Certainly, this is a desirable state to remain in, and a condition you would probably like to preserve around your family, your home, your car, and your possessions, but considering the risks and the statistics, that's easier said than done. Accidents, fires, thefts, and burglaries are chief enemies of people and property.

This chapter contains a number of novel projects that will enable you to protect a medicine chest or gun cabinet from an inquisitive toddler, install a burglar alarm in your motel room in just a few seconds, scare away a car thief, without draining your car's battery, and much more.

Regardless of how you use these projects, you'll find them interesting and worthwhile, and they will help you appreciate the many ways electronics can be of use in gaining the elusive state—*safety!*

"Electric eye" devices have been with us for many years. Although they are commonly used to actuate mechanisms such as doors and counters, they can also be used in burglar-alarm applications. All of the various arrangements in use today are two-part systems. They consist of a strong light source and a photocell unit, both of which must be accurately aligned for proper operation. In security applications, electric eyes are frequently obvious to even the untrained observer, and they generally protect only a limited "line of sight."

As shown in Fig. 38-1, the Watchful Eye is a completely self-contained unit housed in an unobtrusive plastic icebox container. It operates from *any available source of light.*

The sensitivity of the unit increases as the ambient light level decreases. A street light shining through a window provides enough light to operate the unit, allowing it to detect an intruder moving up to 10 feet in front of its "Cyclopean eye."

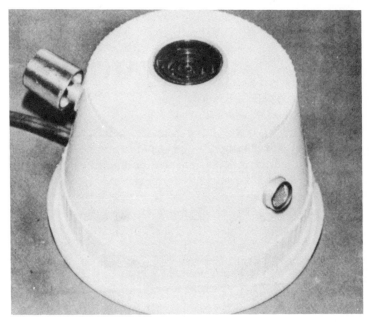

Fig. 38-1. The Watchful Eye.

How It Works

The Watchful Eye alarm is an extremely sensitive light-interruption detector. Its excellent sensitivity is obtained by using a silicon controlled rectifier as a regenerative amplifier, rather than by using it as a switch. When an SCR is operating at very low current levels, it can be considered a complex amplifier made up of two transistors with their inputs and outputs connected for high regenerative feedback. When viewing the SCR in this sense, it must be considered that the device can have enormous gain. This is the principle underlying the unique sensitivity of the Watchful Eye alarm.

As shown in the schematic (Fig. 38-2), the SCR is operated from the ac line. Diode D1 in series with relay RLY1 produces slight positive pulses each time the line swings positive. These pulses are applied to the anode of the SCR. A network consisting of resistors R1, R2, and capacitor C2 couples a portion of the pulses appearing at the anode of the SCR to potentiometer R4. The setting of this potentiometer determines the amplitude of the positive pulses fed to the gate of the SCR. The 4.7 k resistor, R3, in series with the potentiometer limits the maximum amount of pulse excitation that can be applied to the gate. Shunting R3 and R4 is photoresistor R6. This device is a cadmium selenide photoconductive cell that has the property of changing its electrical resistance in proportion to the amount of light that strikes its sensitive surface. The characteristic of the device is such that its resistance is high when little or no light strikes the cell and the resistance decreases as the light increases. (At 2 footcandles, the cell resistance is approximately 1500 ohms.)

In operation, sensitivity control R4 is set so that the SCR

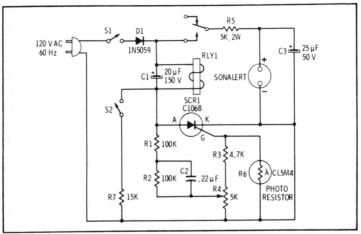

Fig. 38-2. Watchful Eye schematic.

receives almost, but not quite, enough amplitude of the positive pulses to cause regenerative action, and thereby pass sufficient current flow to actuate relay RLY1. This setting is done with the photoresistor exposed to normal ambient light. In this way, the Watchful Eye can be set at any room lighting conditions, whether bright, moderate, or relatively dim. The control is normally adjusted until the unit just sounds off and then the control is backed down slightly until the unit silences. Now the Watchful Eye is armed. If the unit cannot be silenced at any setting of R4, too much light is reaching the photocell and a shield should be used or the illumination level should be reduced.

After the unit is armed, the slightest reduction in the amount of light striking the photoresistor, even on a short-term basis (such as a wave of the hand or the brief passage of a body in front of the photocell), will block some of the light and thereby cause the photoresistor to increase its resistance. As this occurs, higher-level pulses are permitted to appear at the gate of the SCR, which now commences regenerative amplification. In essence, what has happened is that sufficient input voltage has been applied to the SCR gate to make the SCR draw enough current to close the relay. As the contacts of the relay close, the Sonalert is energized by a positive voltage obtained via rectifier diode D1 from the ac line. This voltage is dropped to approximately 20 volts by resistor R5 and is filtered to dc by capacitor C3. The Sonalert used here is a pulsating-sound type. It generates a 2500-Hz warning sound when voltage is applied, and this sound is modulated (pulsed) at a rate of 3 to 5 pulses per second. The result is a very startling sound, unlike any other sound you are likely to hear. When S2 is in the momentary position, the SCR will be triggered on for only as long as the photoresistor senses a change in light conditions, caused by movement in front of the cell. The SCR shuts off after this period, since it is being used at such a low current level that latching action does not occur. When S2 is closed, it is in the latch position. Now it is possible to make the relay latch or remain closed after the light striking the photocell has changed only once. This is accomplished by R7 which places a constant drain of current through the relay winding. This resistor allows a certain amount of current to flow through the relay, which by itself is not sufficient to close the relay. Once the relay is closed by actuation of the SCR, the current flow through R7 is sufficient to "hold" the relay closed. In this case, once the unit has been triggered and has latched, it is necessary to turn off switch S1 to allow the relay to drop out.

Table 38-1. Watchful Eye Parts List

D1—general-purpose silicon diode, type 1N5059 or equiv
C1—20 μF, 150-Vdc electrolytic capacitor
C2—.22-μF, 100-Vdc capacitor
C3—25-μF, 50-Vdc electrolytic capacitor
SCR1—silicon controlled rectifier (General Electric Type C106B,
    Radio Shack 276-1067, or equiv)
R1, R2—100,000-ohm, ½-W ± 10% resistor
R3—4700-ohm, ½-W ± 10% resistor
R4—5000-ohm linear potentiometer (P. R. Mallory type U14 or equiv)
R5—5000-ohm, 2-W ± 10% resistor
R6—photoresistor (Clairex type CL5M4)
R7—15,000-ohm, ½-W ± 10% resistor
Sonalert—P. R. Mallory type SC-628P or Radio Shack 273-060
S1—P. R. Mallory type US26 (part of R4)
S2—spst slide switch
Misc—plastic container, perf board, power cord, wire

Construction

Refer to Table 38-1 for the parts list. The Sonalert, R4, R3, C3, the photocell, and switch S1 are housed in the container as shown in

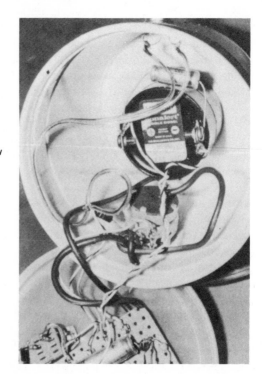

Fig. 38-3. Internal view of the container.

Fig. 38-3. The remainder of the circuit is mounted in the lid of the container, which forms the bottom of the completed unit. As shown in Fig. 38-4, most of the circuit is constructed on a small piece of perf board. The perf board is then attached to the container lid with bolts and nuts. Switch S2 is also mounted on the lid. Note that the lid is slightly recessed. This provides clearance for the switch and the nuts used to mount the perf board.

Fig. 38-4. Perf board and switch mounted on container lid.

39. MEDICINE CABINET ALARM

A youngster's curiosity and agility combined with the accessibility of medicines, household cleaning products, or your gun cabinet, could result in tragedy. Innocent children constantly explore and get into places they shouldn't. That's why it's important that you know what your child is getting into—before it gets into him or her.

The most common causes of childhood poisoning are ordinary household products, as well as the drugs that we all have in our medicine chests. Aspirin and children's flavored medicines are, of course, responsible for a high percentage of oral poisonings, probably due to the fact that all too often we call this medicine "candy" to induce children to take it. Not far behind in lethality are daddy's guns and power tools. They look like great toys, but they are deadly to the innocent and uninitiated.

Of course, you could padlock everything in sight, but how can you live in a home where everything is locked up? (Even more perplexing, how can you be sure to remember to lock up everything, when you may need access to it several times each day?)

The medicine cabinet alarm shown in Fig. 39-1 provides an answer. It can be used wherever you wish, so that you will know when someone, who has no business doing so, has opened a door. Place an alarm inside each cabinet, case, or closet in which hazardous materials are stored. It gives you free access, but it will "squeal" on the kids if they go exploring.

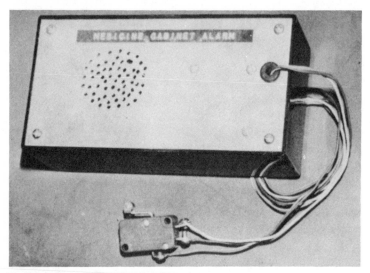

Fig. 39-1. The Medicine Cabinet Alarm.

Here is what happens: When a child opens the door, he has only eight seconds to intrude, after which the alarm will sound and take the child's attention away from getting into trouble. Mom or dad comes running and "saves" the child and everyone is happy.

But now, let us assume that mom or dad comes to the medicine chest during the night to take an aspirin for a headache. The medicine chest is opened and the medication is taken out. If the adult feels that more than eight seconds will be needed to close the door again, he simply holds the microswitch door sensor closed to deactivate the circuit and prevent the alarm from sounding. When he is through, he closes the door in silence.

Thus, the uniqueness of the medicine cabinet alarm lies in the time delay. While any switch, battery, and buzzer combination can be hooked up to blast off if someone opens a cabinet door, it would be most annoying to be startled awake in the middle of the night just because your mate had a headache.

How It Works

The circuit, shown in Fig. 39-2, uses a roller-type spdt microswitch (S1) to sense whether the door is open or closed. This switch (Fig. 39-3) has a normal, no-pressure position and a spring-loaded position, requiring that pressure be applied to the roller. In our application, switch S1 is mounted so that the latched cabinet or closet door applies pressure to the roller, compressing the switch spring. In this closed-door position, the internal battery applies a

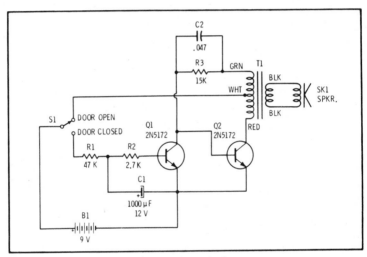

Fig. 39-2. Medicine Cabinet Alarm schematic diagram.

Fig. 39-3. Switch used with Medicine Cabinet Alarm.

charge to capacitor C1, through resistor R1. The capacitor's voltage appears across the base-emitter junction of transistor Q1, through R2. However, neither transistor Q1 nor Q2 has collector voltage, since the switch contacts providing a path to the battery are open when the door is closed. If the door is now opened, switch S1 immediately springs to its no-pressure position, supplying collector voltage to Q1 and Q2. However, the charge stored in capacitor C1 instantly forward-biases Q1 through resistor R1. Transistor Q1 saturates, pulling down the base voltage of Q2 through R3 and C2. This upsets the biasing of Q2, so that the stage cannot oscillate. However, as the charge on capacitor C1 bleeds off after several seconds, Q1 ceases to conduct and the bias rises on the base of Q2 until its gain and the positive feedback through transformer T1 are sufficient to drive the stage into audio oscillation. A speaker connected to the secondary winding of T1 then converts the oscillator's output into a raucous alarm tone that is loud enough and distinctive enough to be heard anywhere throughout the house.

In the interval before the oscillator sounds, the cycle can be cancelled by moving switch S1 back to the door-closed position.

Construction

The complete alarm is housed in a 5-inch × 2½-inch × 1½-inch Bakelite case with a metal cover (see Table 39-1). The loudspeaker is glued onto the cover, behind a number of small holes, which are drilled to allow the alarm sound to come through. As shown in Fig. 39-4, all of the electronic components are mounted on one side of a perf board, with the exception of the transformer, which is secured

Table 39-1. Medicine Cabinet Alarm Parts List.

R1—47,000-ohm, ½-W ± 10% resistor
R2—2700-ohm, ½-W ± 10% resistor
R3—15,000-ohm, ½-W ± 10% resistor
C1—1000-μF, 25-Vdc electrolytic capacitor
Q1, Q2—npn transistor, type 2N5172 or equiv (Radio Shack 276-2009)
T1—transistor transformer; Pri: 1000-ohm ct, Sec: 8-ohm (Argonne type AR-137, Radio Shack 273-1380, or equiv)
S1—spdt roller-leaf type snap switch (Microswitch type 111-SM2-T or equiv)
B1—9-V battery (Mallory MN1604 or equiv)
SK1—2″ diameter, 8-ohm speaker
Misc—5″ × 2½″ × 1½″ plastic case with aluminum panel (Radio Shack 270-233), 2″ × 3″ perf board, battery connector, solder, wire

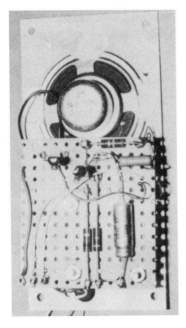

Fig. 39-4. Internal view of Medicine Cabinet Alarm showing component layout.

to the other side. The 9-volt transistor radio battery, B1, is held in place by three screws that also hold the perf board to the cabinet cover. The three leads to the microswitch go through a grommet as shown in Fig. 39-5. We made these leads 1½ feet long, which is adequate for an average-size medicine cabinet. The alarm rests on a

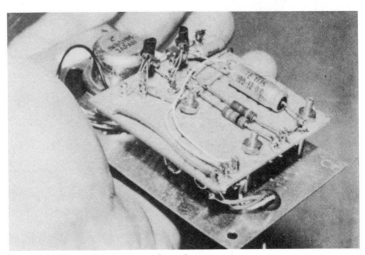

Fig. 39-5. Microswitch leads go through grommet.

shelf in the cabinet, and switch S1 is glued or screwed to the side of the cabinet, so that when the door closes, the switch is actuated.

The time delay is directly proportional to the value of capacitor C1. A 1000-μF value gives about 8 seconds of delay. To get 5 seconds, use 500 μF; to get about 15 seconds, use 2000 μF, and so forth.

40. PORTABLE SECURITY ALARM

When you travel, you probably can't afford to hire your own personal security guard, but if you'd still like a measure of security while you sleep, you can build the electronic Security Alarm shown in Fig. 40-1.

The Portable Security Alarm is a little cylindrical device that measures no more than 2 inches in diameter and about 6 inches in length. It sounds an alarm if someone enters your room while you sleep—and only you can shut it off.

Fig. 40-1. The Portable Security Alarm.

How the Alarm Is Used

The alarm is just the size for packing in your suitcase. Once you arrive at your destination, set it up in your hotel or motel room to guard your belongings against intruders (Fig. 40-2). You can also stand it on top of your suitcase when you must duck into a phone booth to make a call; you will be warned at once if someone attempts a snatch.

Aside from the doorway use pictured in Fig. 40-2, you can set the unit on a windowsill or on the floor in the middle of a hall—

Fig. 40-2. Placing the alarm by a door so that it will be tipped over by anyone who tries to enter.

anywhere an intruder might inadvertently knock it over in the dark. By attaching a simple trip wire to the unit, you could protect a fairly large area.

When triggered by an intruder, the unit emits a loud, continuous series of audible beeps. The alarm will sound whenever it is jarred, knocked over, or moved from the position in which you left it. Once it is triggered, the alarm cannot be turned off by anyone who is not aware of the secret way to deactivate it.

How It Works

The alarm signal is produced by a pulsating Sonalert, activated through a silicon controlled rectifier and powered by two 9-volt alkaline batteries connected in series. The schematic for the circuit is shown in Fig. 40-3. The intrusion sensor used in the security alarm is a mercury switch. This type of switch consists of two normally open contacts that are closed when a pool of liquid mercury comes between them. In this case, a momentary contact is sufficient to close the switch and trigger the SCR.

Once the SCR is turned on, it "latches on" and applies power to the Sonalert. The SCR will not normally turn off unless it is commutated. This is accomplished in our alarm by the use of a Darlington-amplifier touch switch.

Because the Darlington amplifier is such a sensitive device, very little current is required at its base to turn it on. That is the secret of the security alarm. The connection from the base of the Darlington to the plus terminal of the battery is made through your fingertips.

By holding the unit as shown in Fig. 40-1, an infinitesimal amount of current is supplied to the base; the device turns on and commutates the SCR. To turn the alarm off, you must hold it in an upright or other predetermined position, and the travel switch (S1) must be on. No sneak thief will hang around for very long with an alarm sounding off that he cannot silence.

The travel (on/off) switch is needed so that the device can be kept in a deactivated state during transportation. Once the alarm is put into the proper spot and the switch is turned on, the alarm is armed. Any jarring or movement will turn the SCR on and start the blasts. Now, if somebody picks up the alarm, they will naturally want to turn it off by sliding the travel switch to the off position. Once they do this, they've had it! No matter what is tried, this device can never be turned off with the switch in the off position. Only after the alarm is properly silenced do you slide the travel switch to the off position so that you can move the security alarm any way you like.

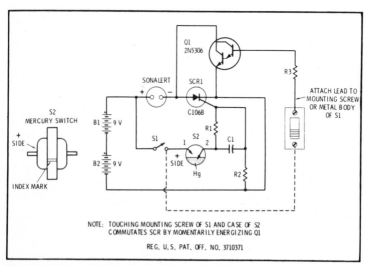

Fig. 40-3. Schematic diagram for the Portable Security Alarm.

Construction

The unit is housed in a frozen-fruit-juice container, which can be covered with some appropriate contact paper and capped on both ends by 2-inch-diameter plastic spray-can tops. One of the tops has a hole for mounting the Sonalert. The one on the bottom raises the center of gravity and provides a slightly rounded bottom to make the security alarm easier to tip over. See Table 40-1 for a complete parts list.

Table 40-1. Portable Security Alarm Parts List.

C1—0.1-μF, 80-Vdc capacitor (Sprague type 192P1049R8 or equiv)
R1—1-megohm, ½-W ± 10% resistor
R2—1000-ohm, ½-W ± 10% resistor
R3—39,000-ohm, ½-W ± 10% resistor
S1—spst miniature slide switch (Radio Shack 275-406)
S2—mercury element from GE mercury toggle switch or Radio Shack 275-027
Q1—Darlington transistor (General Electric type 2N5306 or equiv)
B1, B2—9-V alkaline batteries (Mallory type MN-1604 or equiv)
Sonalert—P. R. Mallory type SC628P or Radio Shack 273-060
SCR1—silicon controlled rectifier (General Electric type C1 6Y1, Radio Shack 276-1067, or equiv)
Misc—perf board, 2″ dia plastic spray-can tops, 2″ dia × 3½″ frozen juice can, wire, solder, bolts, nuts

The mercury element can be taken from any mercury wall switch, available in most hardware stores. The switch is taken apart and the mercury element is removed. This element is mounted between two small perf boards, and contact is made through the heads of two binding-head screws that fit into the small concave spaces on each face of the mercury element. In Fig. 40-4, these screws are seated and wired as they will appear after assembly of the boards. The heads are not attached to the element, which turns freely between them. The two perf boards are spaced approximately ⅝ inch apart and are held in place by three screws that exert the proper pressure on the mercury-switch element.

Two openings are cut in the side of the container. One allows the mercury-switch element to protrude in order to adjust its position. The second opening is for the slide switch. The housing of the slide switch is connected to the Darlington amplifier as shown in the schematic (Fig. 40-3).

Frozen-fruit-juice cans are usually lined with a metal foil. In order to be sure that there are no shorts, place some insulating tape

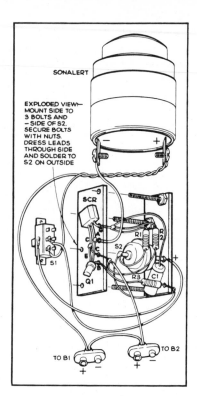

Fig. 40-4. Exploded diagram showing assembly details of the portable security alarm.

on the inside of the container. The two batteries serve to hold the perf boards in place, making for a very compact assembly. Note that the batteries are placed one-half way up in the container, giving a high center of gravity for easy tipping.

41. POWER-LINE BROWNOUT MONITOR

When a brownout occurs, your lights go dim, the picture on your television screen shrinks, and your electric coffee percolator takes a little longer than usual to perk. You tend to shrug—it's just the power company cutting back the voltage a bit to conserve juice on a hot day. No harm done. Other things are happening, however, that can cause harm.

Effects of Voltage Drop

Each lamp, heating device, motor, or other piece of electrical equipment is designed to operate at a certain voltage. Therefore, its efficiency and performance are adversely affected if operated at a lower voltage.

In terms of actual *damage*, resistive loads such as light bulbs or electrical heaters are really not affected. In fact, incandescent lamps last much longer when operated at reduced voltages, although they give substantially less light under those conditions. If the voltage applied to an incandescent lamp is decreased, there is a resulting change in life, efficiency, current, power, and light output.

A 5-percent reduction in voltage will give a 16-percent reduction in light output, and a 195-percent increase in lamp life. A 10-percent reduction in voltage will give a 30-percent reduction in light output and a 393-percent increase in lamp life. *A voltage drop of only 15-percent cuts light output almost in half!*

Mercury-vapor lamps with nonregulated ballasts are affected about the same as incandescent lamps. However, fluorescent lamps and mercury-vapor lamps with constant-voltage ballasts are less affected by voltage. A 5-percent reduced voltage gives 3-percent less light output. However, mercury-vapor and fluorescent lamps will not light at all without enough voltage to ionize the gas in the lamps for starting. A 10-percent reduction in voltage may cause problems with some lamps, especially when the lamp is old. Although most of these lamps will start with a 5-percent reduction in voltage, there will be a 9.75-percent reduction in light output. A 10-percent reduction in voltage will give a 19-percent reduction in light output, and a 20-percent drop in voltage will result in a 35-percent drop in light output.

For example, in an electrically heated building a 500-watt baseboard heater operated at 10 percent below the rated voltage would draw only 405 watts. It would take longer to heat the room and on a cold day might not be able to do the job. If the design were based on heating the room to 70°F with an outside temperature of 0°F, then the room temperature would not get above 57°F when it was 0°F outside.

Television sets are especially susceptible to lower voltages. Voltage reduction can often result in a longer warmup time and a poor picture, especially on a color set.

So far so good. Although we get less light and less heat, there is, as yet, no damage. In fact, as mentioned earlier, the life of resistive heaters and incandescent lamps is prolonged by operation below their "rated voltage."

This is not true for induction motors. They must operate at or near their rated voltage, or they are in for trouble. A drop of only 10 percent in line voltage can cause running torque to decrease 10 percent, slippage to increase 23 percent, current draw to jump 11

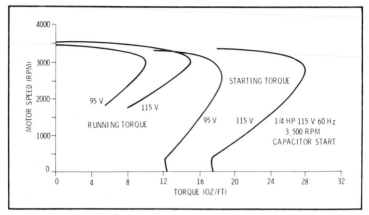

Fig. 41-1. Torque/speed curve for a two-pole induction motor.

percent, and temperature to rise 12° to 15°F. The curve in Fig. 41-1 shows the effect of voltage reduction on the torque of a typical two-pole induction motor. When the voltage applied to an induction motor is reduced below the voltage shown on the nameplate, the efficiency drops rapidly. Also, the current rises drastically for a relatively small reduction in applied voltage. These effects are shown by the curves in Fig. 41-2.

Also, when the applied voltage is reduced 10 percent below the rated voltage, the maximum overload capacity decreases 10 percent. What happens is that the motor, struggling to operate on the reduced line voltage, draws excessive current in an attempt to

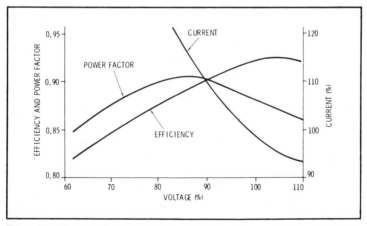

Fig. 41-2. Effect of reduced voltage on efficiency and current draw of induction motor.

compensate for the loss of power. If the line voltage drops too low, intense heat builds up and the motor can burn out. This is an especially serious problem with high-current-drain appliances like air conditioners, but it can also affect power tools and other motor-driven home equipment. Some motors have built-in-overload protection, but many do not and their loss can mean a costly repair or replacement job.

While power companies don't intentionally allow line voltages to fall below the safe limit, accidental brownouts can occur. The power-line brownout monitor shown in Fig. 41-3 keeps an eye on your line voltage and automatically beeps a loud warning if the voltage falls below a preset level. It also sounds an alarm if the voltage goes off altogether, as in the case of a power failure. While a blackout is obvious if it occurs at night when you are awake, it may not be if you are asleep or if it happens during the day when lights are normally off. The monitor alerts you that there has been a power failure and keeps you from being fooled by clocks that don't tell the right time and alarm clocks that don't go off when they should. The power-line brownout monitor includes a standard 0-1 milliammeter, a Mallory Sonalert that sounds a high-pitched beep when energized, and a sensing circuit that monitors line voltage and triggers the alarm at a preset voltage. The parts fit neatly into a 6¼-inch × 3¾-inch × 2-inch Bakelite case (Fig. 41-4). While any suitable

Fig. 41-3. Power-Line Brownout Monitor.

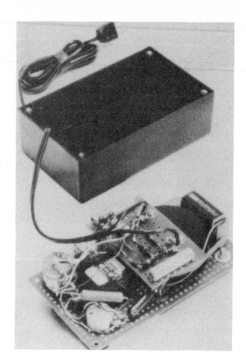

Fig. 41-4. Internal view of the Power-Line Brownout Monitor.

enclosure will do, this type was selected because it has a removable perf-board front panel that is easily cut and drilled for installing the meter and Sonalert. The volt adjust potentiometer (R2), the trip potentiometer (R7, and the set/monitor switch (SW2) also mount in this panel.

The small parts—one SCR, three diodes, two capacitors, and five resistors—mount on a small, separate piece of perf board measuring 2⅞ inches × 2⅛ inches. This board is held in place by the two terminal screws on the back of the meter, as shown in Fig. 41-5. A conversion scale is needed for the meter so that it reads in volts instead of milliamperes. The one shown in Fig. 41-6 is designed to fit a standard 3⅛-inch-square panel meter. Cut out the scale in Fig. 41-7 and paste it over the existing meter scale, being very careful not to touch the delicate meter needle. Wire the unit as shown in the schematic diagram (Fig. 41-8). Power resistors (R1 and R9) should be mounted in free space as they get hot and need good ventilation.

To use the power-line brownout monitor, simply plug its line cord into any convenient wall outlet. With the set/monitor switch in the monitor position (see schematic diagram at SW2), the meter reads ac line voltage as it appears at the outlet. To set the monitor to

218

Fig. 41-5. Component layout for the Power-Line Brownout Monitor.

the desired trip point, flip the switch to the set position. This cuts in the volt adjust potentiometer (R2). Adjust this control until the meter needle reads the voltage level at which you want the alarm to sound. Since 115 to 120 volts is normal in most areas, a setting of around 110 volts provides a safe trip point.

With the set/monitor switch still at set, adjust the trip potentiometer (R7) to the point where the Sonalert just goes on, then flip the switch back to monitor and leave it there during use. The

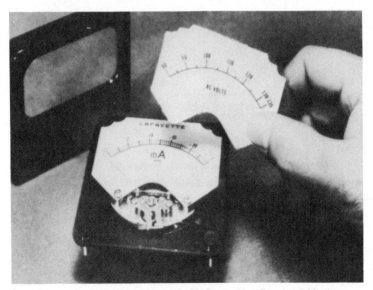

Fig. 41-6. Replace original meter scale for Power-Line Brownout Monitor.

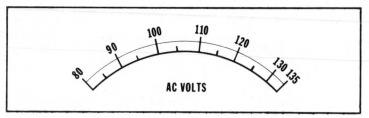

Fig. 41-7. Actual-size meter scale for Power-Line Brownout Monitor.

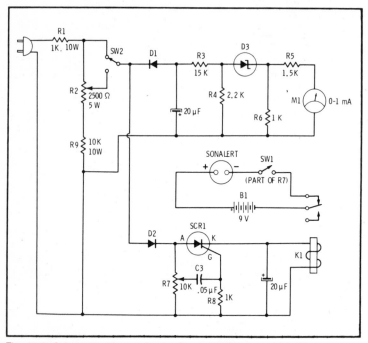

Fig. 41-8. Schematic diagram for the Power-Line Brownout Monitor.

Sonalert will now sound off whenever voltage drops to the preset level. As soon as it does, shut off your air conditioners and other heavy-current-drain equipment until line voltage returns to normal as indicated by the meter. The Sonalert can be temporarily silenced by turning the trip control fully counterclockwise, opening switch SW1. When power returns to normal, reset the trip control to its original position. The Sonalert is powered by a 9-volt transistor-radio battery so that it is independent of house current. Hold the battery in place by fastening it to the meter housing with a heavy rubber band.

A complete parts list is given in Table 41-1.

Table 41-1. Power-Line Brownout Monitor Parts List.

D1, D2—400-piv, 1-A general-purpose silicon diode
D3—10-V, ½-W zener diode (Motorola HEP Z0220 or equiv)
SCR1—silicon controlled rectifier (General Electric C103B,
   Radio Shack 276-1067, or equiv)
C1, C2—20-µF, 150-Vdc electrolytic capacitor
C3—.047-µF, 200-Vdc Mylar capacitor
R1—1000-ohm, 10-W wirewound resistor
R2—2500-ohm, 5-W wirewound potentiometer (P. R. Mallory
   type VW2P5K or equiv)
R3—15,000-ohm, ½-W ± 5% resistor
R4—2200-ohm, ½-W ± 5% resistor
R5—1500-ohm, ½-W ± 5% resistor
R6, R8—1000-ohm, ½-W ± 5% resistor
R7—10,000-ohm, ½-W linear taper potentiometer (P. R.
   Mallory type U20 or equiv)
R9—10,000-ohm, 10-W wirewound resistor
M1—0-1 milliammeter (Lafayette Radio 99R50403, Radio
   Shack 270-1752, or equiv)
SW1—spst switch, part of R7 (P. R. Mallory type US26 or
   equiv)
SW2—spdt miniature toggle switch
K1—midget relay, spdt contacts, 5000-ohm coil, 1.4-mA
   pull-in, 1.2-mA drop-out (Lafayette Radio 30F63005 or equiv)
Sonalert—P. R. Mallory type SC-628 or Radio Shack
   273-060
Misc—6¼″ × 3¾″ × 2″ Bakelite case with perf-board
   front panel, 9-V transistor battery, battery connector,
   perf board, line cord, hookup wire, solder

42. HEADLIGHT-ACTIVATED SAFETY FLASHER

If you have to leave your disabled car to seek help in the dead of night, this dual-purpose Safety Flasher (Fig. 42-1) will lie dormant until it is illuminated by the headlights of an oncoming vehicle. It will then suddenly spring to life and begin to flash its warning message as long as the photocell receives some light. When it is dark again, the flashing stops—and so does the drain on your battery! This means that running out of gas won't mean running out of juice to start your car when you return. In an alternate mode, flicking the switch from automatic to continuous shunts the photo-cell and the flashing is continuous, regardless of the light level.

The circuit, shown in Fig. 42-2, consists of a silicon controlled rectifier (SCR1) and a self-flashing bulb connected across the car's 12-volt dc supply. The photocell is a cadmium sulfide photoconductive cell whose resistance is high in darkness, but drops sharply when it is struck by light. When the sum of resistance R1 and the resistance of the photocell is low enough to trigger the gate of the

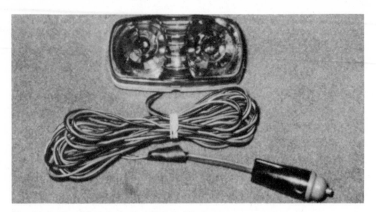

Fig. 42-1. Headlight-Activated Safety Flasher.

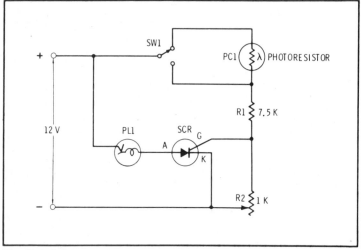

Fig. 42-2. Schematic diagram for the Headlight-Activated Safety Flasher.

SCR, it will begin to conduct and the bulb will start to flash. It will continue to flash on and off as long as the photocell is illuminated by some light. When it is dark again, the photocell's resistance will increase and the lamp will stop flashing. It will remain off until the SCR is turned on again by illuminating the photocell or until the photocell is shunted by switch SW1. The light level at which the safety flasher springs into action is controlled by R2.

The unit is constructed in a *Signalstat* No. 1211 (or similar type) lamp housing, which is a two-lens assembly (Fig. 42-3). One side is used for the photocell and the other for the self-flashing bulb as shown in Fig. 42-4. The slide switch that selects automatic

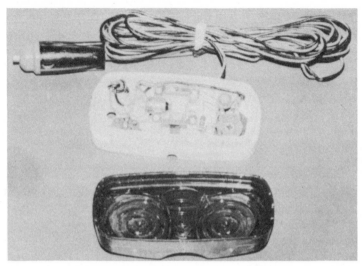

Fig. 42-3. Headlight-Activated Safety Flasher with lens removed.

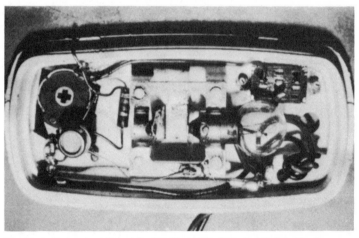

Fig. 42-4. Component layout for the Headlight-Activated Safety Flasher.

operation or continuous operation is mounted in the back of the housing.

The flasher can be mounted in the car's rear window and be permanently connected to the battery through an on-off switch. Or, it can be operated from the cigarette lighter through an adapter available for that purpose.

The parts list for the Headlight-Activated Safety Flasher is given in Table 42-1.

Table 42-1. Headlight-Activated Safety Flasher Parts List.

PC1—Photoconductive cell (Clairex type CL705L)
PL1—flasher lamp (General Electric type 257)
R1—7500-ohm, ½-W ±10% resistor (Use lower value
    for increased sensitivity)
R2—1000-ohm, ¼-W subminiature control (P. R. Mallory
    MTC13L1 or equiv)
SCR1—silicon controlled rectifier (General Electric type
    C106Y1 or equiv)
SW1—spst slide switch
Misc—clearance lamp assembly (Signalstat type 1211 or
    equiv, available from most automotive parts houses)

43. DOOR-CHIME BURGLAR ALARM

Installing a burglar alarm can be a tough job if you have to start from scratch. If you have an electric door chime, with doorbell buttons at each entry to your home, you can expand its function to include after-hours protection of your home and loved ones against a break-in by a thief. In the process, you'll save yourself the cost of an alarm sounder, wiring, and an ac supply, because the door chime circuit already supplies these. The Door-Chime Burglar Alarm is shown in Fig. 43-1.

Here is how the circuit (Fig. 43-2) works: A magnetic reed switch is installed on the door jamb and its mating magnet fits on the door. When the door is closed, the magnet's field holds the reed switch contacts closed. Opening the door swings the magnet out of the proximity of the reed switch and the reeds open.

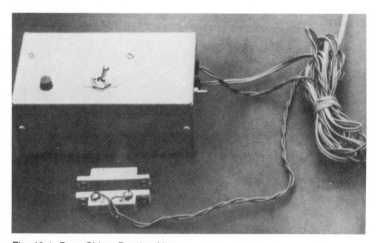

Fig. 43-1. Door-Chime Burglar Alarm.

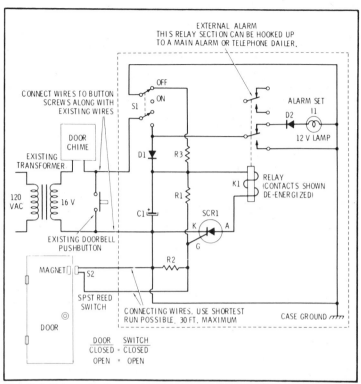

Fig. 43-2. Schematic diagram for Door-Chime Burglar Alarm.

When the reeds open, a trigger voltage appears across resistor R2 at the gate of SCR1. This voltage triggers SCR1 into conduction, energizing relay K1 from the dc supply consisting of diode D1 and filter capacitor C1. The ac power for this supply is continuously provided from the door-chime transformer by connecting leads across the doorbell push button. In this way, enough power is delivered to D1 and C1 to supply operating current to the alarm circuit. However, the current flow is well below the level required to actuate the door chime. Because the SCR is conducting a direct current, it latches in the "on" state, holding relay K1 in. Immediately, the contacts of K1 close the door-chime push-button circuit and the chime sounds its loud warning.

Meanwhile, the ac supply voltage across diode D1 and capacitor C1 drops to zero. The charge stored in C1 is dissipated through relay K1 and SCR1 until the current flow decreases to the point where K1 drops out. (SCR1 remains conductively latched because its holding current is less than 1 mA, well below the drop-out current of relay K1.) As K1 opens, the chime sounds again, but current now flows through diode D1 to recharge capacitor C1 and another cycle commences.

Thus, once the circuit has been triggered, the chime continues to sound in "bing-bong" fashion, providing a loud warning that an intruder has entered your home. The only way to silence the chime is to turn alarm switch S1 off. Resistor R3 is provided to discharge capacitor K1 when the alarm is switched off, preventing false activation of SCR1 by residual charge.

So that you will know that the alarm is set, diode D2 provides rectified half-cycle power to lamp I1 when switch S1 is in the on position and relay K1 is not activated. This helpful reminder will save you the embarrassment of accidentally triggering the alarm if you have to go outside after it's been set.

Construction

The parts list is given in Table 43-1. All components fit neatly inside a 7-inch × 5-inch × 3-inch miniature aluminum minibox, as shown in Fig. 43-3. The relay (K1) is mounted on the end of the minibox, while lamp I1, switch S1, and a 7-terminal tie strip mount on the front face. All wiring is point to point, and no wire lengths are critical.

Table 43-1. Door Chime Burglar Alarm Parts List

```
C1—1000-μF, 35-Vdc electrolytic capacitor
D1, D2—50-piv, 1-A general-purpose silicon diode
I1—12-V indicator lamp (Industrial Device type
    2990D1-12V, Radio Shack 272-322, or equiv)
K1—12-Vdc relay, dpdt contacts (Guardian type
    905-2C-12D or equiv)
S1—dpdt toggle switch
S2—magnetic reed switch for door or window (closed circuit)
SCR1—silicon controlled rectifier, 30-piv, 800-mA (Motorola
    HEP R1001, Radio Shack 276-1067, or equiv)
R1—4600-ohm, ½-W ±10% resistor
R2—1000-ohm, ½-W ±10% resistor
R3—15-ohm, ½-W ±10% resistor
Misc—5¼″ × 3″ × 2⅛″ aluminum minibox
    (Bud type CU-2106A, Radio Shack 270-238, or equiv),
    screws, nuts, wire, solder
```

Installation and Use

Several protection schemes are possible with the door chime burglar alarm. This version protects a single door, but it is possible

Fig. 43-3. Component layout for the Door-Chime Burglar Alarm.

to expand its coverage to include additional doors. Just wire the closed-circuit switches protecting each door and window in series, to form a protection "loop." Then, connect the two ends of the loop in place of the wires leading to switch S2 on the schematic. In this way, the opening of any switch in the loop will immediately raise the voltage on the gate of SCR1 above ground and cause the chime alarm to sound. There is one limitation, however. With the component values given, wire runs should be limited to 30 feet to prevent stray pickup voltages from triggering the SCR. Where a long run is unavoidable, it may be necessary to experimentally reduce the values of R1 and R2 to obtain satisfactory results.

Any type of switch device you wish can be used to protect a door or window. The magnetic proximity switch is a natural choice because it mounts easily and accommodates almost any door or window. Figure 43-4 shows the correct installation for a door.

Fig. 43-4. Magnetic reed switch installed at the top of a door.

227

Tanks, pipes, boilers, water heaters, furnace water lines, and boat hulls all have one thing in common: *they can leak!* Though mending any of these can be a chore, the damage that leaking water can do adds injury to the insult of the repair job. The best solution is to catch the leak while it's still a trickle, and that calls for early warning.

The inexpensive little water alarm shown in Fig. 44-1 is so easy to build that you may want to put together several. Self-contained and battery-powered, it will constantly monitor any area for the presence of water, assuring you by its silence that dry conditions prevail. Should water bridge its sensor probe tips, however, the Water Alarm will sound off to let you know that it's time for action! The penetrating 2500-hertz tone of the Sonalert is distinctive and far-reaching, and the unit is rugged enough to withstand humid or salt-air atmospheres that might silence other noise sources in bonds of corrosion. Add to this the fact that there is no electric spark in the sensor or Sonalert (thereby preventing ignition of gas fumes), and you have an alarm unit that's hard to beat for reliability, simplicity, and low cost.

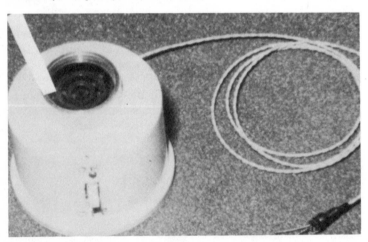

Fig. 44-1. The Water Alarm.

How It Works

Figure 44-2 is a schematic diagram of the Water Alarm. The circuit consists of a monolithic Darlington transistor (Q1), which acts as a high-sensitivity switch for control of current through the

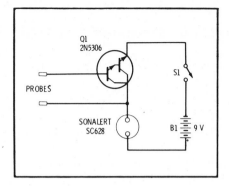

Fig. 44-2. Schematic diagram for the Water Alarm.

Sonalert from a 9-volt battery. The gain of Q1 is so high that a current of a few microamperes through the base-emitter junction is sufficient to drive the stage into saturation, thereby switching on power to the Sonalert. Conversely, in the absence of current flow through the base-emitter junction of Q1, the device blocks current flow through the Sonalert, drawing no power from the battery.

Transistor Q1 is physically arranged as shown in Fig. 44-3 so that its base and collector leads are jutting outward, to act as probes. The collector normally is at the +9-volt level of the battery, while the base is at zero level. This is the normal condition when the probes are dry in the air.

Now, let us assume that water bridges the probes (transistor leads). A slight amount of natural salts and minerals exists in all

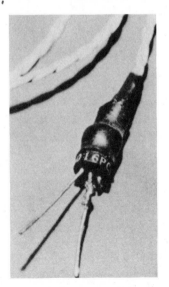

Fig. 44-3. Physical arrangement of the transistor.

water, and this makes it a fair electrical conductor. As the water surrounds both probes, current flows between collector and emitter, energizing the Sonalert. Sound is produced as long as water lies between the probes of transistor Q1, or until switch S1 is opened to break the battery path.

Construction

The Sonalert, Switch S1, and battery B1 fit nicely into the discarded snap-on top of a spray can (Fig. 44-4). Flexible wires run from the components inside the top and are soldered to the collector and emitter leads of transistor Q1. The emitter lead of Q1 is then bent back up over the body so that it will not be able to contact water at the same time as the base and collector leads. If you wish, insulating sleeving, shrink tubing, or epoxy can then be applied over Q1 to insulate everything but the probe leads.

Placing transistor Q1 at the end of a wire run allows you to mount the alarm unit where its sound can be directed toward listeners. This is advantageous because the places where water leaks usually occur are rarely the places where an alarm's sound would be audible. You can use No. 22 insulated stranded wire, up to 25 feet long, for this purpose.

Construction details within the top are obvious from Fig. 44-4 and can be adapted to suit the housing you elect for use.

A complete parts list for the Water Alarm is given in Table 44-1.

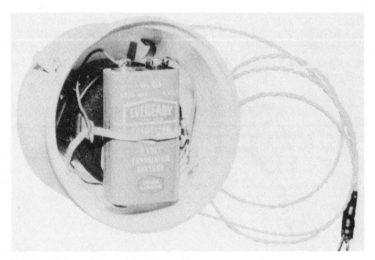

Fig. 44-4. Internal view of the Water Alarm.

Table 44-1. Water Alarm Parts List.

B1—9-V battery (Eveready type 216 or equiv)
Q1—npn Darlington transistor, type 2N5306 or equiv
S1—spst slide switch
Sonalert—P. R. Mallory type SC628 or Radio Shack 273-060
Misc—snap-on top from an aerosol can, wire, solder

Using the Water Alarm

The tiny "sensor," Q1, is adaptable to mounting wherever you need to detect the presence of water. A plastic cable clamp supplies a neat method of mounting Q1 on a baseboard and ensures that the probes will remain oriented toward the surface where water might accumulate. In a boat bilge, you might wish to install the sensor at a low point, such as a convenient strut or brace. You can use epoxy or cyanoacrylate adhesive to bond the sensor to metals, plastics, or masonry. Just be sure that the probes are where you want them, normally in air, and correctly positioned to sense water that accumulates.

The alarm unit can be fastened to any surface, though it's best to choose a low-temperature area for longer battery life. If the top you have chosen comes from a spent aerosol can, you may wish to cut the flange away from the can and devise a way to secure this to the wall or ceiling. This will produce a nifty snap-on mount for the alarm unit, making battery changing quite easy.

You may also wish to experiment with different probe arrangements than the basic one described. Just about any conductive material can be used as a probe, allowing you to adapt the sensor's capabilities to fit your needs. Rigid wire, window screen, or tv-antenna twin lead (with bared wire ends)—all are good possibilities. Simply connect the probes to the base and collector leads, respectively, of sensor transistor Q1.

45. IC DELAYED HEADLIGHT SHUTOFF

You pull into your garage, turn off the engine and lights, and close the garage door. Enveloped in total darkness, you stumble over the garbage cans, almost break your neck on your kids' skates, and then fumble for the keyhole in the door to the house.

Has this happened too many times? Probably, and we would guess that every time it did you wished your parking lights or headlights would stay on for another minute or so after you had gotten out of the car, so that you could safely find your way out of the garage's hazardous environment. This can now be done with our Delayed Headlight Shutoff shown in Fig. 45-1. After a predetermined time, it will automatically turn your headlights off after you have safely entered the house. What's more, the glow of the headlights provides a strong deterrent against muggers and rapists who find a dark garage or driveway such an appealing place to launch an attack.

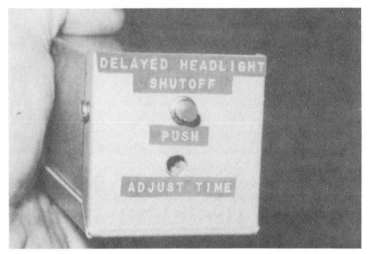

Fig. 45-1. Delayed Headlight Shutoff.

How It Works

The Delayed Headlight Shutoff consists of a 555 timer integrated circuit (IC1) which controls the state of the transistor switch Q1 and relay K1 (Fig. 45-1). The normally open contacts of the relay are simply wired in parallel with the car headlight switch. Therefore, closure of the relay contacts will supply power to the headlights, even though the light switch may be off.

Integrated circuit IC1 is connected as a monostable multivibrator. Its *on* period is determined by the time constant of R1, R2, and C1. Resistor R2 can be adjusted for periods ranging up to 3 minutes. This provides ample time to light your way to the cozy confines of your home while also giving you enough range for several trips back and forth to the car to unload groceries or luggage.

Operation of the circuit is simple. With both the engine and the headlight switches off, momentarily press switch S1. This applies a

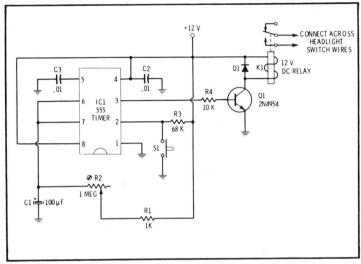

Fig. 45-2. Schematic diagram for the Delayed Headlight Shutoff.

ground pulse to pin 2 of IC1. Instantly, the IC "flips" to its timing state, and the output at pin 3 jumps from ground to +12 volts, biasing transistor Q1 on and energizing the coil of relay K1. The relay pulls in and its normally open contacts close, supplying power to the headlights. Meanwhile, capacitor C1 begins charging to the supply voltage through resistors R1 and R2. The time it takes to charge C1 is the period for which the relay remains energized.

When capacitor C1 has reached supply voltage, a comparator inside IC1 "flips" the circuit back to its stable state. The output at pin 3 of IC1 drops to ground, transistor Q1 turns off, and relay K1 is de-energized. This causes the contacts of K1 to open, and the headlights go out, since the headlight switch is off. The circuit draws infinitesimal current (a few microamperes) in the off state.

To protect the IC from transients, capacitors C2 and C3 bypass the 12-volt supply line and internal IC connection, respectively. Diode D1 limits the inductive spike transients appearing across the coil of relay K1 when power is switched off at the end of a timing cycle. Resistor R1 allows pin 2 to "sense" +12 volts during off periods. It also permits this point to be safely grounded to start a timing cycle, by limiting the amount of current that can flow when switch S1 grounds pin 2. Resistor R4 limits base current into Q1 when the IC is in a timing cycle.

Because the circuit is stable in the off state, draws negligible current between timing cycles, and cannot compromise the normal operation of the headlight switch, there is no need for a separate on-off switch.

Construction

The socket for the 555 timer IC as well as all other components (see Table 45-1) with the exception of the relay are mounted on a small piece of perf board that measures 1½ inches × 2 inches. The board itself is held in place by four ⅝-inch long 6-32 binding-head screws (Fig. 45-3). Four ¼-inch spacers provide the necessary clearance to prevent the circuit from shorting to the case. Time-adjust potentiometer R2 is mounted vertically on the perf board in such a way that its screwdriver-adjustable shaft is located directly behind a small hole in the 2¼-inch × 2¼-inch front face of the aluminum case that houses the unit. The small push button that starts the timing cycle is located directly above as shown in Fig. 45-1. The relay is mounted behind the perf board. We found it convenient to wire diode D1 directly across the coil terminals of relay K1. The actual physical layout may vary somewhat depending on the relay that you use.

The assembled board and relay fit neatly into a miniature aluminum case measuring 4 inches × 2¼ inches × 2¼ inches, although you can gather these parts together in an even smaller enclosure. The leads run directly from the circuitry within the box to the car wiring. The +12-volt and ground leads can be ordinary insulated hookup wire. However, the two leads connecting the relay

Table 45-1. Delayed Headlight Shutoff Parts List.

C1—50-μF, 12-Vdc electrolytic capacitor
C2, C3—.01-μF disc or Mylar capacitor
R1—1000-ohm, ½-W ±10% resistor
R2—1-megohm subminiature control (P. R. Mallory type MTC16L1 or equiv)
R3—68,000-ohm, ½-W ±10% resistor
R4—10,000-ohm, ½-W ±10% resistor
S1—spst normally open push-button switch (Radio Shack 275-618)
D1—1-A, 200-piv silicon diode (Motorola HEP R0052 or equiv)
Q1—npn transistor; 2N4954 or equiv (Radio Shack 276-2009)
K1—12-Vdc relay with 10-A contacts (Guardian type 915-2C-12D, Radio Shack 275-218, or equiv)
IC1—type 555 integrated circuit (Radio Shack 276-1723)
Misc—4″ × 2¼″ × 2¼″ aluminum minibox (Bud type CU-2103A or equiv), dual in-line 8-pin IC socket, 16-gauge stranded insulated wire, hookup wire, solder, etc.

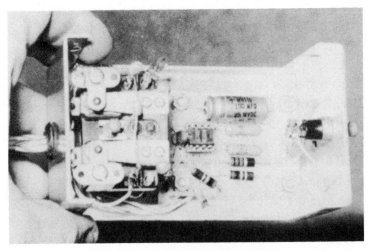

Fig. 45-3. Internal view of the Delayed Headlight Shutoff.

contacts to the headlight switch should be at least 16-gauge standard wire, to prevent excessive voltage drop.

46. THREE-WIRE CIRCUIT MONITOR

The true wiring polarity and integrity of an ac power outlet are not detectable at a glance, but they are far too important to take for granted when you consider the risks involved. The Three-Wire Circuit Monitor shown in Fig. 46-1 can quickly determine the condition of an outlet. If you own a boat and plug into shore power at the marina, or if you are on a trip and connect your trailer or camper power cord into a local outlet, do you know *for sure* that the supply outlet is correctly wired and has a good ground? If you use ac outdoors, for jobs around the house or to power swimming-pool accessories, are you certain that the wiring is safe and has correct polarity? How about home and shop ac outlets? And don't forget extension cords.

Numerous accidents occur because the all-important ground connection is poor or open. (Aluminum wire in contact with other metals can and *does* electrolyze, causing quiet, insidious failure at screw-terminal connections.)* A faulty tool or appliance plugged

---

* To overcome this problem, Underwriters Laboratories imposed stringent standards on wiring devices to be used with aluminum wire after September 1971. If your home is wired with aluminum wire and was built before September 1971, a careful check is in order.

235

Fig. 46-1. The Three-Wire Circuit Monitor.

into a faulty outlet poses a real hazard to the user, since the "hot" side of the line may contact the metal housing or case and find a lethal path to ground through the user's body.

How It Works

The three-wire circuit monitor consists of a flexible cord and plug, leading into a case housing a 0-150 Vac voltmeter and three neon indicator lamps. The schematic diagram is shown in Fig. 46-2. The plug and cord set let you insert the plug into outlets behind furniture or other obstacles and to keep the unit out in the clear, where the meter and indicators can be conveniently read. The plug (Fig. 46-3) is the three-prong type that includes a ground wire.

The meter is wired between the hot and neutral wires of the cord set and will read the rms ac voltage between these two wires when the cord is plugged into an outlet. The neon indicators used

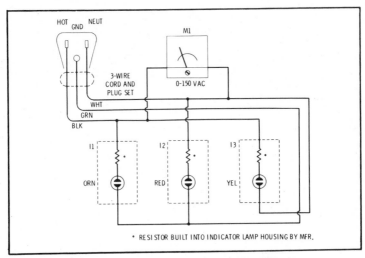

Fig. 46-2. Schematic diagram for the Three-Wire Circuit Monitor.

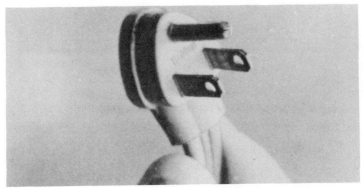

Fig. 46-3. Three-prong plug used with the Three-Wire Circuit Monitor.

contain built-in current-limiting resistors and are wired as follows: lamp I1 is connected between hot and ground; lamp I2 is connected between neutral and ground; and lamp I3 is connected between hot and neutral.

Because the neutral and ground wires are essentially at the same potential in a properly wired installation, plugging in the unit should cause lamps I1 and I3 to glow. The meter should read the proper line voltage (minimum 105-volts ac; maximum 125-volts ac; average value in the U.S.: 117-volts ac). Lamp I2 will not glow because the ground and neutral wires are essentially at identical potential.

Where an open-hot-wire fault has occurred or a circuit breaker has tripped, the meter will read 0 volts and none of the indicators will glow.

A reversal of polarity (wrong wires on wrong terminals) has many possible results. These are listed in Table 46-1.

Table 46-1. Faults and Indications.

Condition	Meter	Lamps
Properly wired outlet	Reads normal line voltage (U.S. average 117 Vac)	11○ 13○ 12●
Open ground	Reads normal line voltage	11● 13○ 12●
Open hot wire	Reads zero	11● 13● 12●
Open neutral wire	Reads zero	11○ 13● 12●
Hot-neutral reversed	Reads normal line voltage	11● 13○ 12○
Hot-ground reversed	Reads zero	11○ 13● 12○

Lamp extinguished ●
Lamp glowing ○

An open ground connection is exceedingly treacherous because the outlet remains capable of supplying the voltage to a device, although it gives no warning that the lifesaving ground connection no longer extends to the case or housing of the device being used. In this instance, the circuit monitor meter will read line voltage, and only indicator lamp I3 will glow.

An open neutral wire will, in a properly installed outlet, remove line voltage from the outlet. This will be evidenced by a zero-volts meter reading and only lamp I1 will glow (assuming that the ground wire is intact).

Construction

The checker is housed in a Bakelite instrument case for complete insulation of the device from the user's body. Holes for the indicator lamps are drilled and carefully reamed to size to avoid cracking the panel. The meter hole can be made by drilling a circular series of small-diameter holes to form the perimeter, then chipping out the material between the holes to remove the center. Wiring is simplified by using a wire nut to splice the indicator-lamp leads to

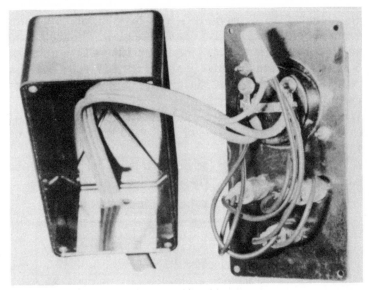
Fig. 46-4. Internal view of the Three-Wire Circuit Monitor.

the ground wire. Other leads connect to the two meter terminals, as shown in Fig. 46-4.

The parts list is given in Table 46-2.

Table 46-2. Three-Wire Circuit Monitor Parts List.

I1—orange 120-V neon indicator lamp assembly
I2—red 120-V neon indicator lamp assembly
I3—yellow 120-V neon indicator lamp assembly
M1—0-150 Vac voltmeter (Calectro type D1-926 or equiv)
Misc—6″ × 3″ × 2¼″ Bakelite case with phenolic cover,
    three-wire line cord with grounding-type plug, wire nut, solder, etc.

47. AUTO-ALARM

Here is an unusual car alarm (Fig. 47-1) that gives no outward sign of its presence but is a real "time bomb" for any unwary thief. Imagine this scene: The car thief, using a master key, pick, or "slapper," opens the door of your car. He slips inside, smiling confidently, little realizing what's about to happen! Just as he is about to start the car, the horn starts blowing *and won't stop!* Now, if there's one thing a car thief does not want, it's the attention of a crowd! So, he bolts from the car and is out of sight within seconds.

Fig. 47-1. The Auto-Alarm.

The horn continues sounding for another 90 seconds, then automatically silences as the alarm resets itself to save your car battery and your neighbors' nerves from further abuse.

Even more attractive than the invisibility of this unique alarm is its simplicity of installation. Just three connections are necessary to install the alarm: a wire to a +12-volt line, a wire to the car chassis or ground, and a connection to the wire running from the horn switch to the horn relay. There is no need to install sensing switches to trip the alarm—the circuit is designed to detect *any* current drain from the car battery, thus making every electrical device in the car an extension of the alarm system! Thus, when a door is opened and the dome light comes on, the alarm will be tripped. If the ignition is switched on, the alarm will be tripped. The same for the headlights, electric windshield wipers, heater, radio *anything* that draws current from the battery.

To give *you* time to disarm the alarm after entering the vehicle, a 15-second delay is provided from the time the alarm is tripped until the horn sounds. This gives you time to flip a hidden switch that deactivates the alarm.

How It Works

The circuit for the Auto-Alarm and its connection to the car wiring are shown in Fig. 47-2. Assuming that switch S1 is in the off position, the driver prepares to leave the car by opening the door, so that the dome light goes on. Next, switch S1 is moved to the on position and the door is closed and locked. (Opening the door *before*

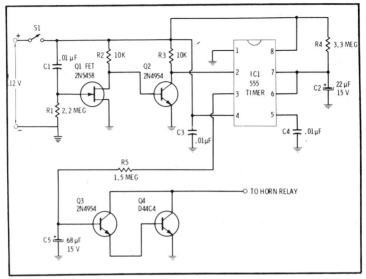

Fig. 47-2. Schematic diagram for the Auto-Alarm.

arming the alarm is imperative, since it is the inital surge of current to a load which triggers the circuit.)

When the circuit is armed but not yet triggered, field-effect transistor Q1 is coupled to the 12-volt supply line through capacitor C1, with gate return supplied by resistor R1. The conducting drain-source channel of Q1 holds the base of a transistor Q2 at near ground potential so that Q2 is in a nonconducting state. Thus, there is essentially no voltage drop across resistor R3, and +12 volts is applied to input pin (2) of timer IC1. In this stable state, the output pin (3) of IC1 is at ground potential, thus holding the Darlington delay circuit, consisting of transistors Q3 and Q4, in the nonconducting state.

However, when the door is next opened, the courtesy light switch connects the dome light across the battery. This produces a discernible negative-going spike on the 12-volt line, owing to the fact that the lamp has exceedingly low cold resistance. This spike is sensed by the circuit in the following manner.

At the instant the lamp is connected, the 12-volt line instantaneously drops in potential so that capacitor C1 attempts to discharge. This causes a negative potential to appear across resistor R1, briefly biasing the gate negative so that drain current ceases. Immediately, base current flows into transistor Q2, biasing it on and driving the collector voltage close to ground potential. The steep

241

fall of the Q2 collector voltage at pin 2 of IC1 triggers the time delay circuit on. Even though transistors, Q1 and Q2 recover almost immediately, the timer IC is enabled and must complete its cycle.

When the IC is triggered, pins 6 and 7 are no longer at ground potential, and the capacitor C2 immediately commences to charge toward the supply voltage through resistor R4. The values of these components are chosen so that it takes about 2 minutes to charge C2. Meanwhile, at the instant of triggering, the output pin (3) of IC1 rises from ground to +12 volts. This sets in motion another time delay in the base circuit of Q3 and Q4. Since the IC1 output remains "high" throughout the initial 2-minute timing cycle, capacitor C5 begins charging to the supply voltage through resistor R5. The values C5 and R5 are chosen so that this takes about 10 seconds. At the expiration of this time, Q3 and Q4 conduct, providing a ground return for the car-horn relay.

The relay closes and the horn sounds. The horn remains on, so long as the IC1 output supplies enabling base current to Q3 and Q4. Upon expiration of its 2-minute cycle, the output of IC1 drops to ground, removing turn-on bias from Q3 and Q4. This causes the circuit to revert to its stable state, awaiting another triggering impulse. (Capacitors C3 and C4 decouple the 12-volt line supply to the IC to prevent false triggering.)

Let us now suppose, however, that it is you who is entering the car. From the time of entry, IC1 has been triggered. You now have about 15 seconds to turn off the alarm. Flicking the hidden switch to off disarms the circuit, resets IC1, and makes sure that the horn doesn't blow.

Construction

A parts list for the Auto-Alarm is given in Table 47-1. Parts arrangement is not critical. The parts can be installed on perf board and placed in a small plastic box to insulate connections and shut out dust. A terminal board simplifies hookup of the wires that you install in the car. Fig. 47-3 clearly shows parts placement and layout.

Recommendations

If you have hood and trunk lights of the sort containing mercury switches, these two vital areas will also be protected against violation by a thief or "stripper." (A hood lock is also an effective first line of defense, and your car should have one.) If your car lacks these features, consider their addition for extra protection.

If your car has an electric windup clock, you may have to

Table 47-1. Auto-Alarm Parts List

C1, C3, C4—.01-μF, 200-Vdc Mylar capacitor
C2—22-μF, 15-Vdc tantalum capacitor
C5—68-μF, 15-Vdc tantalum capacitor
IC1—type 555 integrated circuit (Radio Shack 276-1723)
Q1—n-channel field-effect transistor, type 2N5458 or equiv
Q2, Q3—npn transistor, type 2N4954 or equiv (Radio Shack
   276-2009)
Q4—npn power transistor (General Electric type D44C4
   or equiv)
R1—2.2-megohm, ½-W ±10% resistor
R2, R3—10,000-ohm, ½-W ±10% resistor
R4—3.3-megohm, ½-W ±10% resistor
R5—1.5-megohm, ½-W ±10% resistor
S1—spst switch (type determined by choice of location
   in the car)
Misc—perf board, plastic box, wire, solder, etc.

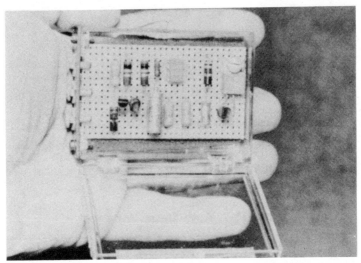

Fig. 47-3. Component layout for the Auto-Alarm.

disconnect it. These clocks wind by means of a solenoid, actuated every minute or so. Because the solenoid pulses the 12-volt line, it can lead to false alarms. (Based upon our experience with these rarely reliable timepieces, the choice seems to be protecting the car or knowing the *wrong time*. A good wristwatch is all you need to help you make up your mind.)

Your options in placing the hidden alarm switch are many, indeed. You may choose to mount a small toggle switch behind the dash where only you can find it, or you may want to be brazen and mount an automotive-type switch in plain sight on the dash, where it can easily be overlooked as "just another switch." Both types of switches are shown in Fig. 47-4. The choice is yours, and you can use your imagination to the fullest.

Be sure that once the alarm is armed (turned on), you do *nothing* that draws power from the car battery. Open the door *before* you set the alarm, then leave. If you want to be sure that nothing in your car will draw current and falsely trigger the alarm when you're not around, study the car maker's service manual and review your car's equipment complement for hidden devices that consume current.

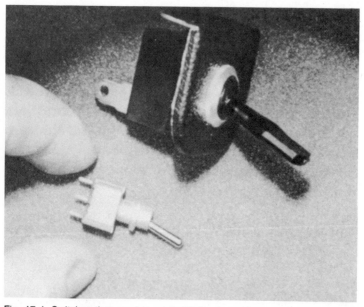

Fig. 47-4. Switches that can be used with the Auto-Alarm.

48. TOUCH-ACTIVATED SWITCH

If you think Aladdin's lamp was talented, take a look at the versatile Touch-Activated Switch shown in Fig. 48-1. It looks like any other three-outlet, surface-mount ac receptacle, but at the gentlest touch, this "electronic genie" springs into action to light a lamp, ring a warning bell, or sound a buzzer. You can use it as an annunciator to tell you that someone has entered your home or place

244

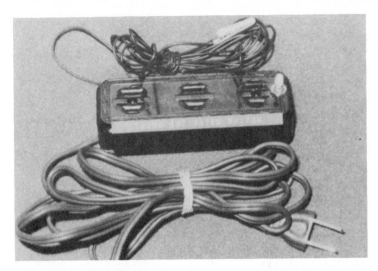

Fig. 48-1. Touch-Activated Switch.

of business, as a "hands-off" device to remind visitors to keep away from things they shouldn't touch. It makes a dandy alarm device to monitor a door or window. It has other uses also; you can use it to turn on a circuit or appliance at the touch of your finger.

Black magic? Nope. The touch-activated switch operates from body capacitance, plus a handful of inexpensive electronic components you build into it. When you touch the "antenna," a tiny alternating current flows through you to ground, your body acting like a small capacitor. The harmless current is so minute you can't feel it yet it is sufficient to trigger the sensitive circuit.

The "antenna" can be any ungrounded metal object—a doorknob, mailbox, trash can, lamp base, touchplate, or simply a wire lead. The circuit will activate any combination of bells, buzzers, lights, or other devices at half the line voltage. However, the total power drawn by them must be under 150 watts.

At the flip of a toggle switch, the device can be converted from a nonlatching to a latching circuit. With the switch open, the device operates only as long as someone is touching the antenna. With the switch closed, a single touch instantaneously triggers the circuit and latches it. The device will keep operating whether or not someone is touching the antenna, until it is manually shut off by opening the switch.

Some capacity-operated devices use oscillators that function when they are detuned by body capacity. These circuits are, however, subject to instability and false triggering, and their sensitivity is not readily adjustable. They also need a relay, and they draw standby power from the power line at all times. The circuit described here needs no relays or other moving parts, and all the electronics can be easily fitted inside a small convenience outlet. As a result, we have a reliable, very inexpensive, and compact little device. Simply plug in what needs to be actuated, and it will go *on* when the "antenna" or the "touch button" is touched. It's as simple as that.

How It Works

The heart of the circuit (Fig. 48-2) is a silicon controlled rectifier. When no voltage is applied to the gate of the SCR, it acts like an open circuit, and no current flows to the load. However, when a positive pulse of sufficient amplitude is applied to the gate of the SCR, anode current flows as long as the anode is positive with respect to the cathode, even if the gate no longer has a potential

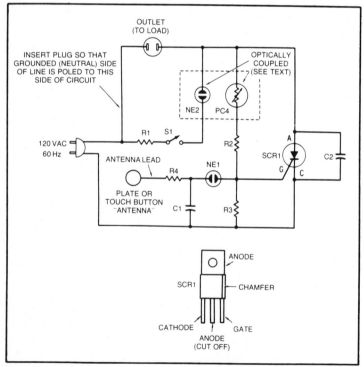

Fig. 48-2. Schematic diagram for Touch-Activated Switch.

applied to it. In order for a positive pulse to reach the gate, neon lamp NE1 must be triggered. If the resistance between the left side of NE1 and the ground side of the line is very high, there is not sufficient voltage developed across NE1 to fire the lamp. However, if a person touches the antenna lead, he effectively lowers this resistance to the point where sufficient voltage is available across capacitor C1 to fire NE1 and send a positive pulse to the gate of the SCR. This causes the SCR to conduct on every positive half cycle of line voltage applied between its anode and cathode.

When switch S1 is closed, PC1, a small cadmium-sulfide photocell, acts as an electronic latch. It permits the SCR to remain conductive after the external capacitance that triggered the gate is removed. When no light is shining on PC1, the resistance of the photocell is so high that it can be considered an open circuit. However, when NE2 fires because the SCR is conducting, the light from NE2 causes the resistance of PC1 to drop to around 130,000 ohms. The resulting low series resistance of PC1 and R2 then biases the SCR into conduction every positive half cycle of the line voltage, removing the need for NE1 to trigger the SCR until switch S1 is opened.

Because SCR1 acts like a rectifier as well as a switch, the current output of the device consists of half-wave rectified dc. In other words, only one-half cycle of the ac input flows through the load. Thus, a lamp plugged into the device will light at less than full brilliance, and a 120-volt ac bell will sound at less than full volume, though still loud enough for most applications.

If you wish to use the touch switch to control other types of light ac leads—a tv or hi-fi, for example—add a power-control relay to the circuit. Any spst or spdt 115-volt relay with contacts rated at three amperes or more will work.

The fact that the voltage is pulsating dc may make it necessary to remove turns from the coils of some 115-volt ac relays. Depending on their design, some ac relays will chatter at the low output voltage of the touch control while others will hold firm. Connect the relay coil to a line cord and plug it into the touch switch-activated outlet; then use the relay contacts to control the load.

Construction

A parts list for the Touch-Activated Switch is given in Table 48-1. The internal parts layout for the Touch-Activated Switch is shown in Fig. 48-3. Use a dark brown, light-tight, plastic receptacle for the housing, so that photocell PC1 is illuminated only by neon

Table 48-1. Touch-Activated Switch Parts List

R1—47,000-ohm, ½-W ± 10% resistor
R2—1-megohm, ½-W ± 10% resistor
R3—27,000-ohm, ½-W ± 10% resistor
R4—100,000-ohm, ½-W ± 10% resistor
C1—50-pF, 500-Vdc silvered-mica capacitor
C2—100-pF, 500-Vdc silvered-mica capacitor
NE1—NE-83 neon lamp
NE2—NE-2 neon lamp
PC1—photocell, Clairex type CL903
SCR1—silicon controlled rectifier (General Electric type C106B1 or
   equiv)
S1—spst toggle switch
Misc—dark-brown receptacle housing, line cord

bulb NE2. Outside light may sabotage the latching action and keep
the SCR turned on at all times.

In order to realize a compact assembly, the anode of the SCR is
directly fastened to the terminal of the outlet by means of the screw
which is normally provided. In order to achieve a good fit, the anode
tab of the SCR has to be trimmed slightly on the side opposite the
chamfer, and the hole in the tab must be somewhat enlarged to
accept the screw that comes with the outlet. The barrier tabs within
the outlet are put to good use by placing one SCR lead on each side,
thus preventing any possible short circuits. Fig. 48-3 clearly shows
how the few parts are put together and inserted inside the plastic
body of the outlet.

A small hole for the antenna wire must be drilled on that side of
the outlet on which the line cord enters. Another hole for the switch
is drilled on the opposite end of the outlet, as shown in Fig. 48-1. No
other physical modifications are necessary.

Tight optical coupling is required between PC1 and NE2. This
is accomplished by wrapping them together with black plastic electrical tape, which makes for a good mechanical fit and provides a

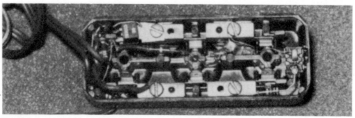

Fig. 48-3. Internal view of the Touch-Activated Switch.

light-tight assembly so that no external extraneous light can fall on PC1 and affect the operation of the device. Keep the antenna lead as short as possible—never over six feet. Although the antenna is shown in Fig. 48-2 as a "touch button," it can be a small button or a large metal plate or wire mesh. As a rule of thumb, the larger the antenna, the more sensitive the circuit. (Be sure that the button or plate used as the antenna is not grounded, or your device will stay on all the time.)

The sensitivity of the device is controlled by the value of capacitor C1, the length of the antenna lead, and the area of the metallic plate (if one is used). The smaller the capacitance of C1, the greater the sensitivity of the unit. A fixed capacitor is used in the unit shown. Its value was determined by adjusting a variable capacitor in a breadboard model for the desired sensitivity. If a variable capacitor is used, a plastic knob or insulated screwdriver should be employed (for safety) to make the adjustment.

Large antennas and long leads may make the circuit unstable. You can reduce the sensitivity by increasing C1 to 100 pF. Conversely, if you are using a very short antenna lead and a small antenna, you can increase sensitivity by decreasing C1 to 25 pF or leaving it out of the circuit.

Note that in order for the touch-operated switch to function, the device must be properly plugged into the line. If the line plug is reversed, there will be no hazard, but the device will not function.

49. ELECTRONIC TEMPERATURE CONTROLLER

If you have a large investment in tropical fish, or if you develop and print your own photos, you know how critical even slight differences in temperature can be to the success or failure of your hobby. Just a few degrees below the right temperature can wreak havoc in an aquarium, or doom your cherished negatives to destruction in the developer bath. Moreover, it's hard to keep track of a freezer's operation when it is a unit located in a basement or garage that you visit infrequently. There, a temperature rise may allow food to spoil, with distressing consequences.

These are just a few examples of applications for the Temperature Controller shown in Fig. 49-1. You can use it to monitor a preset temperature, and either warn you when temperature goes the wrong way, or activate control devices which correct the situation automatically. Basically, the device is a wide-range electronic thermometer, with a go/no-go output. You can set it to monitor

Fig. 49-1. The Electronic Temperature Controller.

temperature well below 0°F and warn you of warm-up. Or, you can set it to monitor a high temperature (up to 300°F) and warn you of cool-down.

The device uses a sealed thermistor sensor and it responds to as little as a 3-percent change in the temperature of a medium. Its operation is based on the principle that a normally hot medium requires action when it begins to cool, and that a normally cold medium requires action when it begins to warm up. As such, the temperature controller is suitable for most one-way control applications. Where a medium must be held to extremely critical temperature, neither rising nor falling more than a few degrees, a more complex controller than this simple circuit may be required.

How It Works

The circuit, shown schematically in Fig. 49-2, consists of glass-bead thermistor sensor R1, control R2, zero-voltage switch IC1, and SCR logic drivers SCR1 and SCR2.

The sensor, R1, is exposed to the medium to be monitored and assumes a resistance proportional to the temperature of the medium. This provides a signal to zero-voltage switch IC1, which can be negated by adjustment of R2 (the reason for this will be explained shortly).

The zero-voltage switch, IC1, is a high-gain amplifier that also has the ability to detect the instant that the applied ac voltage crosses zero, and to deliver an output pulse at that instant. Switch IC1 is thus able to amplify the signal difference between sensor R1

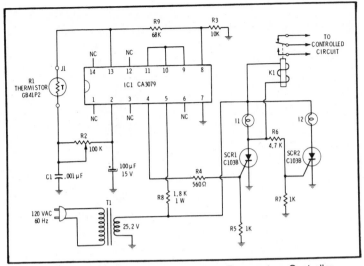

Fig. 49-2. Schematic diagram for the Electronic Temperature Controller.

and control R2 setting, and to deliver a correctly timed pulse to trigger SCR1. In this manner, the circuit responds on each half cycle of the ac line when the difference is sufficient to warrant an alarm or a control response.

Silicon controlled rectifier SCR2 is a "slave" of SCR1. That is, it conducts when SCR1 is not conducting. It is used in conjunction with lamp I2 to indicate that the difference between the R1 absolute resistance (determined by the sensed temperature) and R2 setting is not sufficient to activate SCR1. Silicon controlled rectifier SCR2 receives gate bias through R6 and R7, from the anode of SCR1. When SCR1 is not being triggered by IC1, its anode rises to the ac supply voltage delivered by power transformer T1. Thus, the potential appearing at the junction of R6 and R7 is sufficient to trigger SCR2 on each positive-going half cycle of the supply voltage. As a result, lamp I2 goes on.

However, if control R2 has been set so that there is a significant difference between its resistance and that of the sensor, R1, then IC1 commences to pulse the gate of SCR1 through R4 and R5 on each half cycle of the ac line voltage. Under these conditions, SCR1 turns on and the voltage across it falls to zero, so that gate voltage is insufficient to trigger SCR2. Thus, lamp I1 is illuminated, but lamp I2 is extinguished.

If control R2 has been initially set so that it requires a *decrease* in resistance of sensor R1 to prevent SCR1 from firing, then relay

K1 can be used to energize a heating element. In turn, the heating element heats the medium sensed by R1 and deactivates the element when the resistance of R1 balances the R2 setting. This is one form of feedback that can close the loop between input and output. An aquarium heater or photo darkroom tray heater are possible applications. Of course, the circuit can also be used to activate a warning bell or to detect a rise in air temperature caused by fire. In this case, control R2 is set so that an *increase* in temperature will cause SCR2 to fire. Possible applications are limited only by your ingenuity and your needs.

Construction

The unit is housed in an aluminum minibox measuring 2¼ inches × 2¼ inches × 4 inches. Circuitry and transformer T1 are mounted on perf board (Fig. 49-3), and IC1 is mounted on an IC adapter board (Fig. 49-4) which simplifies connection to its 14 pins. Both boards are mounted by standoff screws. A microphone connector and cable lead to the sensing thermistor, which is epoxied to an insulating probe handle for convenient use, as shown in Fig. 49-5. (You may want to try different mounting methods to serve your purpose and to provide optimum protection for the glass-encased thermistor.) A parts list for the Electronic Temperature Controller is given in Table 49-1.

Table 49-1. Electronic Temperature Controller Parts List

C1—.001-μF, 200-Vdc Mylar capacitor
C2—100-μF, 15-Vdc electrolytic capacitor
IC1—zero-voltage switch integrated circuit (RCA type CA3079)
I1, I2—12-V miniature lamp (Industrial Devices 1090A1-12V or equiv)
J1—2-conductor microphone connector (Amphenol type 75-MC1F or equiv)
P1—microphone connector receptacle (Amphenol type 75-PC1M or equiv)
K1—12-V relay or bell (optional)
R1—glass-encased thermistor 1k @ 25°C (Fenwal GB41P2)
R2—100,000-ohm potentiometer (Centralab type B-40 or equiv)
R3—10,000-ohm, ½-W ± 10% resistor
R4—560-ohm, ½-W ± 10% resistor
R5, R7—1000-ohm, ½-W ± 10% resistor
R6—4700-ohm, ½-W ± 10% resistor
R8—1800-ohm, ½-W ± 10% resistor
R9—68,000-ohm, ½-W ± 10% resistor
T1—power transformer, 24-V @300 mA (Radio Shack 273-1386)
Misc—2¼" × 2¼" × 4" aluminum minibox (Bud CU-2103A or Radio Shack 270-236), perf board, integrated circuit wiring adapter, knob, line cord, nuts and bolts, wire, solder, etc.

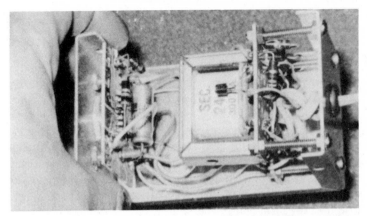

Fig. 49-3. Internal view of the Electronic Temperature Controller.

Fig. 49-4. Internal view of the Electronic Temperature Controller showing IC mounting.

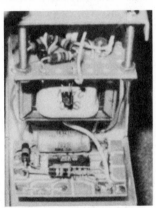

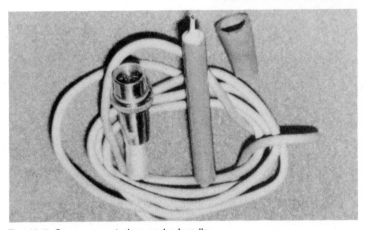

Fig. 49-5. Sensor mounted on probe handle.

It's 3 o'clock in the morning. You are suddenly awake in the unfriendly darkness of your bedroom, listening. What was it that awakened you? The squeak of a window, yielding to a burglar's pry bar? A falling potted plant, kicked from the sill as an intruder clumsily entered your home? You listen, breathlessly, in the silent darkness, waiting for the next sound that will confirm an intruder's presence in your home—and fearing terribly that you will hear it!

Statistics say that a bad economy breeds crime. Burglaries are at an all-time high when need drives individuals who have nothing to lose, to prey upon homeowners. That is why the prudent home-owner should take steps to protect his family and property by installing an intrusion alarm.

The Intrudaflash shown in Fig. 50-1 is a simple, reliable, and economical intrusion alarm that can give you a large measure of protection against the horror scenario just described. It is designed to be activated by passive switch sensors at each door and window, or by manually actuated "panic" button switches placed at strategic points in your home. Once activated, the Intrudaflash will sound a loud bell and turn a bright light on and off in rapid sequence, to summon aid to your home and to put to flight the burglar who has attempted to attack your family and home.

How It Works

Refer to the schematic shown in Fig. 50-2. The instant that contact is established across the sense terminals, which connect to the control winding of the ac-line isolation relay K1, the Intrudaflash goes into action. It begins to apply pulsed 110-V ac power to any combination of lights, bells, sirens, horns, or buzzers plugged into the load outlet, which is mounted on the side of the plastic utility case that houses the Intrudaflash.

At the same time that relay K1 applies 110-Vac (between the blue and the white wires) to that portion of the Intrudaflash circuit which produces the flashing action, it also applies voltage to dc relay K2 so that it closes. Diode D5 rectifies and resistor R1 limits the current through this relay to a safe value. Capacitor C1 provides the filtering necessary to avoid relay chatter. The normally open contacts of relay K2 are connected across the control winding of relay K1 so that they are effectively in parallel with the sense terminals.

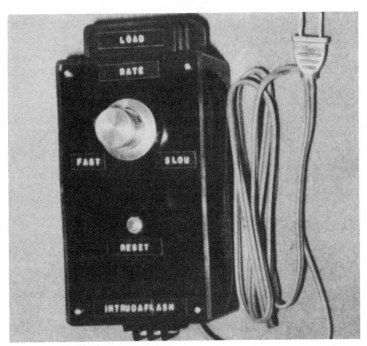

Fig. 50-1. The Intrudaflash.

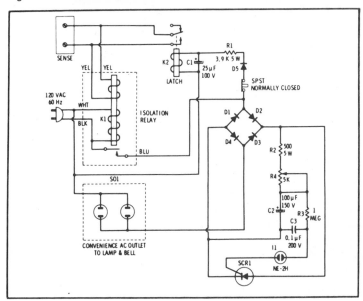

Fig. 50-2. Schematic diagram for the Intrudaflash.

This feedback arrangement latches the unit "on" and guarantees that the flasher circuit continues to operate once it has started, even if the connection across the sense terminals is opened.

The four rectifier diodes, D1, D2, D3, and D4, form a bridge circuit, and the SCR is connected across the dc legs of this bridge. As soon as ac is applied to the bridge circuit, capacitor C2 starts to charge through the series combination of fixed resistor R2 and variable resistor R4. Capacitor C3 now also begins to charge through resistor R3. When the voltage across this capacitor reaches about 90 volts, it fires the neon bulb which then discharges capacitor C3 into the gate of the silicon controlled rectifier. The SCR conducts and applies full line voltage to any load plugged into the load outlet. As soon as the SCR conducts, capacitor C2 starts to discharge through R2, R4, and SCR. For a short time, this action provides a dc current through the SCR which is above its holding current. As a result, the SCR stays on. When the current available from capacitor C2 drops below the holding current, the SCR will turn off during the next ac cycle when the line voltage is near zero. With the SCR off, capacitor C2 starts to charge again through R2 and R4 until the neon bulb fires once more and the cycle repeats. The 5000-ohm potentiometer provides for a variable flashing rate, from 15 flashes per second at the maximum resistance setting of R4, to 80 flashes per second at the minimum resistance setting.

To stop the flashing, simply press the push-button switch. This interrupts the circuit to relay K2 so that it opens. This removes the connection across the control winding of K2 and, unless there is still a connection across the sense terminals, the action stops.

Construction

A parts list is given in Table 50-1. For safety in use, the Intrudaflash is housed in a Bakelite utility case that is closed with a Bakelite cover. The internal construction is shown in Fig. 50-3. All of the component parts, with the exception of isolation relay K1, are mounted on one side of a 3½-inch × 4-inch perf board which is held in place by two machine screws. Isolation relay K1 is mounted next to the perf board and is held in place by two 1¾-inch 6-32 screws. The sense terminal strip, which provides the "input," is located on one side of the case; and the utility outlet, which furnishes the "output," is affixed to the opposite side of the case. The rate adjustment potentiometer and the reset push button are mounted on the Bakelite cover. Be careful to position these two components in such a way that when the cover is screwed in place, neither the

Table 50-1. Intrudaflash Parts List.

C1—20-μF, 100-Vdc electrolytic capacitor
C2—100-μF, 150-Vdc electrolytic capacitor
C3—.1-μF, 200-Vdc Mylar capacitor
D1, D2, D3, D4—3-A, 200-piv general-purpose diode
D5—1-A, 200-piv general-purpose diode
R1—3900-ohm, 5-W resistor
R2—500-ohm, 5-W resistor
R3—1-megohm, ½-W ±10% resistor
R4—5000-ohm potentiometer (P. R. Mallory type U14
   or equiv)
NE—neon lamp type NE-2H (Radio Shack
   272-1102)
K1—isolation relay (Alco type FR105)
K2—miniature relay, 500-ohm coil (Guardian type
   1335-2C-24D or equiv)
SCR1—silicon controlled rectifier (General Electric type
   C-220B or equiv)
SW1—spst normally closed push-button switch (Radio
   Shack 275-1578)
Misc—6¼″ × 3¾″ × 1⅞″ Bakelite case, line cord, duplex
   utility outlet, 3½″ × 4″ perf board, 2-terminal barrier
   strip, knob

Fig. 50-3. Internal view showing parts layout for the Intrudaflash.

potentiometer nor the push button short against any of the components on the perf board or against one of the relays.

Sensors and Wiring

Switch-type sensors should be placed at each door or window you wish to protect. You can use magnetically activated reed switches, microswitches, push-button switches or even "home-brew" switches consisting of a wiping contact mounted on the door and a fixed contact on the doorjamb. All that is necessary is that the switch you choose should be open (no contact) when the door or window is in its closed condition.

For added detection capability, you may want to consider use of pressure-sensitive mats or strips under rugs and carpeting. These are especially good choices for in-house protection, just in case someone should get past the perimeter protection on doors and windows.

It is a good idea to provide one or more "panic" switches at

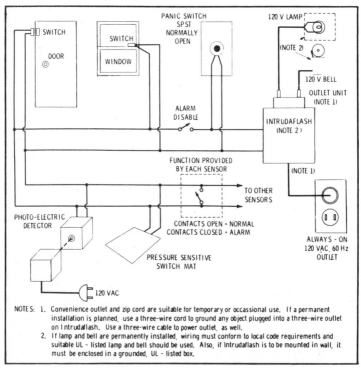

NOTES: 1. Convenience outlet and zip cord are suitable for temporary or occassional use. If a permanent installation is planned, use a three-wire cord to ground any object plugged into a three-wire outlet on Intrudaflash. Use a three-wire cable to power outlet, as well.
2. If lamp and bell are permanently installed, wiring must conform to local code requirements and suitable UL - listed lamp and bell should be used. Also, if Intrudaflash is to be mounted in wall, it must be enclosed in a grounded, UL - listed box.

Fig. 50-4. Typical wiring diagram for alarm system utilizing the Intrudaflash.

strategic points around the home. Consider locations like: beside the bed, in closets next to front and rear doors, in the kitchen or laundry room, etc.

Wiring to the sensors and panic switches consists of two leads that carry a very low voltage, safely isolated from the power line by the transformer action of isolation relay K1. All sensors and switches are simply wired across these two leads so that a momentary closure of any set of sensor contacts will short two wires together.

Figure 50-4 shows a typical wiring diagram for sensors and panic switches and their connection to the Intrudaflash.

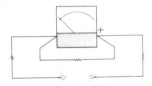

Index